T0094195

New Techniques
for Studying Biomembranes

Methods in Signal Transduction

Series Editors: Joseph Eichberg Jr., Michael X. Zhu, and Harpreet Singh

The overall theme of this series continues to be the presentation of the wealth of up-to-date research methods applied to the many facets of signal transduction. Each volume is assembled by one or more editors who are preeminent in their specialty. In turn, the guiding principle for editors is to recruit chapter authors who will describe procedures and protocols with which they are intimately familiar in a reader friendly format. The intent is to assure that each volume will be of maximum practical value to a broad audience, including students and researchers just entering an area, as well as seasoned investigators.

New Techniques for Studying Biomembranes
Edited by *Qiu-Xing Jiang*

Polycystic Kidney Disease
Jinghua Hu and Yong Yu

Signal Transduction and Smooth Muscle
Mohamed Trebak and Scott Earley

Autophagy and Signaling
Esther Wong

Lipid-Mediated Signaling Transduction, Second Edition
Eric Murphy, Thad Rosenberger, and Mikhail Golovko

Calcium Entry Channels in Non-Excitable Cells
Juliusz Ashot Kozak and James W. Putney, Jr.

Signaling Mechanisms Regulating T Cell Diversity and Function
Jonathan Soboloff and Dietmar J. Kappes

Gap Junction Channels and Hemichannels
Donglin Bai and Juan C. Saez

TRP Channels
Michael Xi Zhu

Cyclic Nucleotide Signaling
Xiaodong Cheng

Lipid-Mediated Signaling
Eric J. Murphy and Thad A. Rosenberger

Signaling by Toll-Like Receptors
Gregory W. Konat

Signal Transduction in the Retina
Steven J. Fliesler and Oleg G. Kisselev

Analysis of Growth Factor Signaling in Embryos
Malcolm Whitman and Amy K. Sater

Calcium Signaling, Second Edition
James W. Putney, Jr.

G Protein-Coupled Receptors: Structure, Function, and Ligand Screening
Tatsuya Haga and Shigeki Takeda

For more information about this series, please visit: https://www.crcpress.com/Methods-in-Signal-Transduction-Series/book-series/CRCMETSIGTRA?page=&order=pubdate&size=12&view=list&status=published, forthcoming

New Techniques for Studying Biomembranes

Edited by
Qiu-Xing Jiang

CRC Press
Taylor & Francis Group
Boca Raton London New York

CRC Press is an imprint of the
Taylor & Francis Group, an **informa** business

First edition published 2020
by CRC Press
6000 Broken Sound Parkway NW, Suite 300, Boca Raton, FL 33487-2742

and by CRC Press
2 Park Square, Milton Park, Abingdon, Oxon, OX14 4RN

CRC Press is an imprint of Taylor & Francis Group, LLC

ISBN: 9781138618060 (hbk)
ISBN: 9780429461385 (ebk)

Typeset in Times
by Lumina Datamatics Limited

Contents

Preface

While working on the structural basis and the energetics of the lipid-dependent gating of voltage-gated potassium channels, a new hypothesis which was raised by my laboratory and has stood the test of time, I realized that the available techniques for studying lipids and membrane proteins in artificial and cell membranes have various limitations. My laboratory has implemented various membrane systems and imaging techniques and developed our own bead-supported unilamellar membrane (bSUM) system. New techniques have been developed by others in different directions in the past decades and could be systematically introduced in new books to researchers in the field and interested readers. Upon the suggestions of Dr. Michael X. Zhu at the University of Texas Health Science Center at Houston, Texas, USA, I took the task of editing this book and a forthcoming book on biological membranes (biomembranes). *New Techniques for Studying Biomembranes* will introduce the latest technologies and the forthcoming book, *Lipid Determanant of Membrane-Protein Interactions* will be focused on the lipid effects on specific membrane proteins. This book is the result of a year or so of preparative work. My initial goal in this part is to introduce newer techniques for analysis of biomembranes. New technologies always drive our studies to a higher level and a newer frontier. Lack of high sensitivity and high spatial and temporal resolutions in studying individual lipid molecules in biomembranes has been a long-standing bottleneck. Different techniques to be introduced in the main body of the book reflect the ample efforts of different investigators in this area. I hope that some of these efforts will ultimately lead to tools for studying individual lipid and protein molecules in a native environment, which once achieved will reveal unprecedented insights into lipid-protein interactions and the co-evolution of membrane proteins with the lipid environments of their native niches.

I would like to take this opportunity to thank Dr. Zhu and all contributing authors. The editorial team at CRC Press, including Dr. Chuck R. Crumly, Mrs. Ana Lucia Eberhart, and their colleagues worked hard to accelerate the publication of this work. I am deeply indebted to their time and efforts. This work is dedicated to Dr. David Gadsby of the Rockefeller University, who passed away early this year and had been a giant in structure–function studies of membrane proteins and whose work has influenced my own work significantly in the past.

Qiu-Xing Jiang
Gainesville, Florida

Editor

Dr. Qiu-Xing Jiang is an associate professor in the Department of Microbiology and Cell Science at the University of Florida, and the head of the Laboratory of Cell Physiology and Molecular Biophysics. In 2020, he is becoming the Director of cryoEM Center at the Hauptman-Woodward Medical Research Institute in Buffalo, New York. He earned a BS in biophysics from University of Science and Technology of China, an MS in cellular biophysics from the Institute of Biophysics at the Chinese Academy of Sciences, and a PhD in cellular and molecular physiology at Yale University. He did a postdoc training at The Rockefeller University before starting his independent research. Afterward, he was an assistant professor of cell biology at UT Southwestern Medical Center in Dallas until he relocated his lab to Florida in 2015. Dr. Jiang received a EUREKA award from NIH of the United States in 2009 and a National Innovative Award from AHA in 2012. His lab has performed substantial amounts of work on membrane biology and molecular biophysics. Currently Dr. Jiang's research programs focus on three general areas: (1) membrane biology with main interests in lipid-dependent gating of voltage-gated ion channels and roles of membrane proteins in the regulated secretory pathway; (2) intracellular RNA-binding complexes and the structural basis for their signaling; and (3) chemical functionalization at the nanometer scale and its application to both cryoEM imaging and biological questions.

Contributors

Dr. Cristina Bertocchi is an assistant professor at the Pontificia Universidad Católica of Chile. Her research interest is to understand the molecular basis and dynamics regulating complex physiological function, in particular how cells communicate, interact and sense mechanical stimuli. She earned her MSc in biology at the University of Milan (Italy) and her PhD in physiology at the University of Innsbruck (Austria). She joined the Mechanobiology Institute (Singapore) as a postdoc, where, working shoulder to shoulder with Prof. Pakorn Kanchanawong, she employed cutting-edge microscopy techniques to investigate the molecular machineries regulating force transduction and mechanosensation at cell-cell adhesion.

Dr. Anna N. Bukiya is an associate professor at the Department of Pharmacology, Addiction Science, and Toxicology, College of Medicine, the University of Tennessee Health Science Center. Her scientific interests lie in the field of lipid modulation of ionic currents and ion channel sensitivity to drugs of abuse starting from *in utero* development and spanning into adulthood. Current studies in Dr. Bukiya's laboratory employ an array of computational, biochemical, electrophysiological, and integrative approaches. Dr. Bukiya is a member of the American Society for Pharmacology and Experimental Therapeutics, Research Society on Alcoholism, and International Drug Abuse Research Society. She is also a current member of the Biophysical Society, in which she serves on a Committee for Inclusion and Diversity. Dr. Bukiya published more than 50 research manuscripts, reviews, and book chapters. She served as an editor/co-editor of several books focusing on cholesterol and cannabinoid modulation of protein function and relevant physiological processes.

Dr. Alex M. Dopico, MD, PhD, is a university distinguished professor and chair of the Department of Pharmacology, Addiction Science, and Toxicology in the College of Medicine at the University of Tennessee Health Science Center. His research career has focused on the modulation of ion channel function by organic ligands of simple chemical structure, whether drugs of abuse (ethanol) or signaling molecules (CO), and by membrane lipids. His multiple original research articles, reviews and book chapters document the regulation of potassium channels of the BK_{Ca} type by fatty acids, leukotrienes, PIP_2 and related phosphoinositides, vasoactive steroids, and cholesterol itself. In 2009, Dr. Dopico received a Merit Award from NIAAA for his work on ethanol regulation of slo channels from arteries and central neurons. He currently serves as a member of the National Advisory Committee on Alcohol Abuse & Alcoholism.

Dr. Timothy J. Garrett is an associate professor in the Department of Pathology at the University of Florida and associate director for the Southeast Center for Integrated Metabolomics (SECIM). His research involves the application of lipidomic and metabolomics from the development of MALDI-based approaches to analyze lipids and small molecules with imaging mass spectrometry to deep metabolome discovery using liquid chromatography-high resolution mass spectrometry. Other current interests are in the application of direct tissue analysis approaches such as MALDI, DESI, and LMJSSP as well as the use of high-resolution mass spectrometry in metabolomics and routine diagnostics. He enjoys the interplay between technological advancement and clinical analysis providing unique opportunities and experiences to develop the future diagnostics.

Dr. Joseph Irudayaraj is a faculty member in the Department of Bioengineering at the University of Illinois at Urbana Champaign. His interests are in developing biosensors and single molecule spectroscopic techniques to examine intracellular protein-DNA interactions, specifically those related to epigenetic regulation. Their group has developed techniques based on SERS for the quantification of biomarkers and intracellular monitoring of biological and chemical modifications. Examples of applications include detection of pathogens, hazardous chemicals, epigenetic modifications, genetic and cell surface markers, and signaling pathways. He teaches courses in the areas of biosensors, bioinstrumentation, and finite element analysis.

Dr. Iqbal Mahmud is a postdoctoral associate of the Department of Pathology, Immunology, and Laboratory Medicine at the University of Florida College of Medicine in Gainesville, Florida, USA. In 2014, Dr. Mahmud earned his MS in biotechnology from Claflin University, Orangeburg, South Carolina, under the supervision of Drs. Kamal Chowdhury and Arezue Boroujerdi, trained on nuclear magnetic resonance (NMR)-based metabolomics and biomarker discovery. In 2018, Dr. Mahmud earned his PhD in medical sciences from the University of Florida, Gainesville under the supervision of Daiqing Liao, PhD, in collaboration with Dr. Timothy J. Garrett and trained on cancer metabolism, epigenetics and therapeutics. Dr. Mahmud's dissertation work first established DAXX (Death Domain Associated Protein) as a critical regulator of oncogenic lipogenesis and a novel therapeutic target for cancer. His current research mainly focuses on the elucidation of metabolic liabilities in cancer and malnutrition in children using different chromatography and high-resolution mass spectrometry techniques.

Mrs. Kelsey North is a PhD candidate at the University of Tennessee Health Science Center currently working in the Department of Pharmacology, Addiction Science, and Toxicology, where she is co-mentored by Dr. Alex Dopico and Dr. Anna Bukiya. She earned her bachelor's degree and master's degree from Austin Peay State University, Clarksville, Tennessee, where she studied hyperpolarization-activated cyclic-nucleotide gated channel regulation of gonadotropin-releasing hormone secretion under the advisement of Dr. Gilbert Pitts. Her current research involves the regulation of cerebral circulation and smooth muscle function by calcium and voltage-gated potassium large conductance (BK) channels. In the three years with Drs. Bukiya and Dopico's group, she published three first-author manuscripts, received three travel awards, presented posters at numerous conferences; all while being an active mentor for her peers as the pharmacology track representative and director of community outreach for the Graduate Student Council.

Dr. Andrea Ravasio is an assistant professor at the Pontificia Universidad Católica de Chile. He earned his master's degree in biology at the University of Milan (Italy) and a PhD at the University of Innsbruck (Austria). With a background in cell physiology and biophysics, his current scientific interests focus on understanding how molecular-scale events in biological systems integrate in space and in time to define cell and tissue function and determine disease states. His technical expertise includes advanced microscopy, bioengineering, and microfabrication. He develops optical methods to access cell-environment interactions and forces therein.

Dr. Lorena Redondo-Morata carried her PhD in the physical chemistry department of the University of Barcelona, concomitantly with the Institute for Bioengineering of Catalonia (IBEC). Her PhD (2012) was devoted to the study of nanomechanics of lipid bilayers mainly using atomic force microscopy (AFM)-based force spectroscopy. In 2013, she joined as a postdoctoral fellow to Dr. Simon Scheuring's laboratory at the Institute de la Santé et la Recherche Médicale (INSERM) in Marseille, France. There, she learned high-speed AFM to study dynamic remodeling of biomembranes. In 2016 she was awarded by the Spanish Biophysical Society a Young Researcher Prize in recognition of her research achievements. In 2018 she obtained an interim tenure researcher position in the Institut Pasteur in Lille, France. At the present, she continues her studies devoted to the mechanics and dynamics of cell membrane remodeling using AFM and correlative microscopies.

Dr. Wen Ren is a postdoctoral associate at the University of Illinois at Urbana-Champaign. He is a nanoparticle chemist and has over a decade of experience in developing nanoparticle biosensors both as beacons for SERS and for colorimetric sensing. His expertise is in nanoparticle chemistry for a range of applications spanning the areas of food safety, environment monitoring, and disease detection. In addition, he is well versed with a range of analytical instrumentation and characterization techniques for examining nanoparticle protein-DNA interactions.

Dr. Yi Ruan earned a BE degree in telecommunication engineering from Beijing Jiaotong University, Beijing, China, in 2007, and a PhD in optics, photonics, and image processing from Ecole Centrale Marseille, Marseille, France, in 2013. His PhD work was devoted to developing far-field super-resolution microscopy, termed as tomographic diffractive microscopy. From 2013 to 2014, he was a postdoctoral researcher at the Photonique, Numérique et Nanosciences Laboratory (LP2N), Centre National de la Recherche Scientifique (CNRS), Bordeaux University 1, Bordeaux, France. There, his studies focused on detecting ultra-high density single molecules interactions in breast cancer by using uPAINT method. From 2014 to 2017 he was a postdoctoral researcher at Dr. Simon Scheuring's laboratory at the Institut National de la Santé et de la Recherche Médicale (INSERM), Aix Marseille University, Marseille, France. There, he learned high-speed AFM to study the dynamic function-related conformational changes of transmembrane proteins in membranes.

He has been an associate professor in Zhejiang University of Technology, Hangzhou, China since August 2017. His research interests include AFM imaging in general and its applications to biology, far-field super-resolution imaging, and inverse scattering problems.

Dr. Timothy J. Rudge earned a bachelor of engineering degree in mechatronics at Leeds University (UK), before studying for a master's degree in computer science at University College London, focusing on image processing and reconstruction. He earned a PhD in biology and carried out postdoctoral research at the Cavendish Laboratory, Cambridge University. He is an assistant professor at Universidad Católica de Chile where he researches the formation of spatial patterns in cell populations. He has extensive research experience from major institutions including Cambridge University and the University of Queensland, (Australia), always with a strong component of microscopy and image analysis.

Dr. Simon Scheuring is a professor of physiology and biophysics in Anesthesiology at Weill Cornell Medicine, New York, USA. He is a trained biologist from the Biozentrum at the University of Basel, Switzerland (1992–1996). During his PhD (1997–2001) in the laboratory of Andreas Engel, he learned electron microscopy and atomic force microscopy and became interested in membrane proteins. During this period he worked on the structure determination of aquaporins and sugar transporters. During his post-doc (2001–2004) and as an Institut National de la Santé et de la Recherche Médicale (INSERM) research assistant (2004–2007) at the Institut Curie in Paris, France, in the laboratory of Jean-Louis Rigaud, he learned membrane physical chemistry and developed atomic force microscopy for the study of native membranes. As a junior research director (2007–2012), he set up his lab at the Institut Curie in Paris, France. Next, he built a larger laboratory at INSERM/Aix-Marseille Université (2012) in Marseille, France, where he was promoted to senior research director (2012–2016). He then moved to Weill Cornell Medicine, New York, USA (2017), where he was appointed as Professor of Physiology and Biophysics in Anesthesiology. Dr. Scheuring has been awarded an INSERM Avenir (2005), the Médaille de la Ville de Paris (2007), a European Research Council (ERC) consolidator grant (2012), the grand prix Robert Debré (2013), and the NIH director's pioneer award (2019). His objective is to head a dynamic research team with members from different scientific fields ranging from biology, physics, and engineering to develop and use atomic force microscopy based technologies to provide novel insights into the processes taking place in biological membranes.

Dr. Liguo Wang earned his PhD from Cornell University, Ithaca, New York in 2003. Then he joined Dr. Fred Sigworth's group at Yale University, New Haven, Connecticut to work on the structure determination of membrane proteins using cryo-Electron Microscopy (cryo-EM). A method termed as the random spherically constrained (RSC) single-particle cryo-EM method was developed and was successfully employed to study the structure of a voltage-gated potassium ion channel. In 2011, he became an assistant professor and started his cryo-EM laboratory in the Department of Biological Structure at the University of Washington, Seattle, Washington. His research was focused on developing the RSC method both experimentally and computationally, and employing the RSC method to determine structures of voltage-gated ion channels in different functional states. In 2019, he moved to the Brookhaven National Laboratory as a cryo-EM group leader.

Dr. Gaya P. Yadav, PhD, is a biological scientist III at the Department of Microbiology and Cell Science, the University of Florida, Gainesville, Florida, USA. In 2020, he is becoming a cryoEM Facility Manager and an Associate Research Scientist at the Hauptman-Woodward Medical Research Institute in Buffalo, New York. His scientific interests lie in the field of lipid-dependent gating of ion channels and ion channels in the regulated secretory pathway. Currently, Dr. Yadav employs an array of computational, biochemical, biophysical, and integrative approaches to study the structure and function of ion channels. Dr. Yadav is a member of the Biophysical Society and American Heart Association. Dr. Yadav published more than 16 research manuscripts, reviews, and book chapters. He served as an associate editor of several research journals focusing on protein function and relevant physiological processes.

1 Introduction
New Tools for Challenges in Membrane Biology

Qiu-Xing Jiang

CONTENTS

1.1 A BRIEF HISTORY OF BIOMEMBRANE STUDIES

Biological membranes (biomembranes) have been the subject of studies for a long time, long before the discovery of DNA double helix (Heimburg, 2007). Since the discovery of cholesterol in blood by M.F. Boudet in 1833, phospholipids by Theodore Gobley in 1847, sphingolipids by J.L.W. Thudichum in 1876, and the coining of the term "lipid" by Gabriel Bertrand in 1923, early studies of biomembranes were limited at the chemical identification of the constituent molecules. Before 1970, a lot was known about the basic properties of biomembranes without the knowledge of the correct arrangements of proteins and lipids in the unit membranes. Low sensitivity of techniques used in biochemical analysis of lipids was rate-limiting.

Technical difficulty also has made discoveries slow-paced in the field. Charles Ernest Overton (1895–1990) proposed that biomembranes are made of lipids, which serve as the basis for differential cell permeability and narcosis, and demonstrated the importance of sodium-potassium ATPase in plasma membranes to muscle and nerve excitability. Agnes Pockels' early design of lipid-spreading on the surface of water was later improved by Irving Langmuir into the so-called Langmuir trough (1917), which was used by Evert Gorter and François Grendel in 1925 to collect data and support their hypothesis of a double layer (bilayer) of lipids in plasma membranes of each red blood cell, which was later firmly established by the EM images of the tri-band structure as described by Walther Stoeckenius in 1962 (Stoeckenius, 1962a, 1962b; Stoeckenius and Engelman, 1969). By then the reconstitution of membrane proteins in vesicles and artificial planar lipid bilayers became feasible. Regarding the protein arrangement in biomembranes, the three-layered protein-membrane-protein

1

model was first proposed by Jim Danielli and Hugh Davson in 1935 and further buttressed by the EM observations of 75-Å wide unit membranes with a 35-Å wide hydrophobic core by J. David Robertson in 1959, but it missed the insertion of membrane proteins in the hydrophobic cores (Danielli and Davson, 1935; Robertson, 1967). The fluid mosaic model proposed in 1972 by S. Jonathan Singer and Garth Nicolson gained favor and popularity in the field because of its timely integration of different lines of evidence and good agreements with experimental observations, especially after incorporating Daniel Branton's results from freeze-fracture electron microscopy that demonstrated the integration of membrane proteins into both leaflets of a cell membrane (Branton, 1966; Singer and Nicolson, 1972). The general ideas of the fluid mosaic model become well accepted that membrane proteins may be attached as peripheral proteins or span one or two leaflets and become integrated (buried) in biomembranes, and that both integral membrane proteins and lipids may freely diffuse laterally within a bilayer membrane. To achieve this general model, biochemical analysis and high-resolution EM studies were both critical in resolving lipid composition of membranes and the relatively uniform dimensions of bilayer arrangements in various types of biomembranes.

In consideration of protein-lipid interactions and phase separations in lipid membranes, modifications of the fluid mosaic model, called modified fluid mosaic model (e.g., Figure 1.1) (Nicolson, 2014), which includes the possible hydrophobic match (or mismatch) between lipids and membrane proteins from the main ideas of the

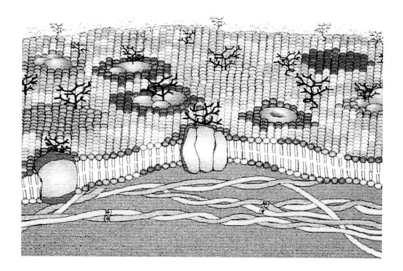

FIGURE 1.1 A schematic presentation of the modified fluid mosaic model of biomembranes. Different types of lipid molecules are colored differently. Cholesterol is not shown differentially. Integral membrane proteins are presented as pink blobs and cytoskeleton filaments are drawn underneath the membrane in the bottom. Glycans on lipids and proteins are showed as arborized sticks. Microdomains are visible. (Reprinted from *Biochim. Biophys. Acta*, 1838, Nicolson, G.L., The Fluid-Mosaic Model of Membrane Structure: Still relevant to understanding the structure, function and dynamics of biological membranes after more than 40 years, 1451–1466, Copyright 2014, with permission from Elsevier.)

mattress model (Mouritsen and Bloom, 1984), lipid rafts and caveolae, become the current generally accepted model for biomembranes, although still qualitative (Jacobson et al., 2007). As shown in Figure 1.1, it is possible that, the first layer of lipid molecules (called the annular layer) around each transmembrane protein have direct interaction with protein residues and would be partially ordered or become less fluidic such that the lipids are partially organized by the protein, forming a protein-centered domain. Atomistic or coarse-grained molecular dynamics simulation of simple phospholipid bilayers has been extensively conducted in studying the energetics and conformational changes of membrane proteins in bilayer membranes with quite remarkable success; for examples, see (Callenberg et al., 2012; Wee et al., 2010), even though the resulted models are still not fully satisfactory in predicting the behavior of native cell membranes that contain high concentrations of membrane proteins and are made of a set of different lipid molecules.

1.2   IMPORTANCE OF LIPID-PROTEIN INTERACTIONS REQUIRES NEW AND BETTER TECHNOLOGIES

For a eukaryotic cell, its plasma membrane separates its interior from the outside, giving the cell its own identity and an open system made of its enclosed interior and its environment to be kept away from thermodynamic equilibrium via constant flow of negative-entropy that is maintained by consumption of chemical energy. Inside the cell, there are membrane-bound organelles that often require ion gradients across membranes for their specialized functions, and membrane-less organelles that form a dynamic mass aggregation via liquid-liquid phase separation at a high protein concentration in order to concentrate functional units to accomplish mass-action-based, highly cooperative activity through entropy-limited thermodynamics process (Bentley et al., 2019; Gomes and Shorter, 2019). Proton-gradient and Ca^{2+}-gradient across membrane-bound organellar membranes and transport of nutrients, metabolites, proteins, and ion across these membranes have been main themes of many interesting studies. So is the lipid distribution and transport within and across biomembranes. Membraneless organelles form through multi-dentate interactions between interacting molecules, domains or motifs that can overcome the negative entropy change in a confined space. Within the membrane-enclosed organelles, highly concentrated proteins, nucleic acids, small biomolecules, etc., are constantly transformed or translocated to accomplish specific biological activities. The relatively high stability of biomembranes is sufficient and important for separating and confining specific molecules in a membrane-bound organelle to fulfill specific physiological functions. An immediate cost of biomembrane-based compartmentalization is that transport of polar materials and specific signals across membranes needs to be mediated by specialized transmembrane complexes that are integrated in bilayer structures and function as structural support, receptors, transporters, channels, etc. These membrane proteins form microdomains with their surrounding lipid molecules and function in a particular niche where a particular lipid composition and a set of protein partners are maintained.

Lipids may exert strong modulatory (energetic) effects on membrane proteins. The importance of lipid molecules to the structure and function of membrane proteins has been a long-pursued question, even though studies of membrane proteins in different membrane systems of different cells tend to focus on the key functional characteristics that remain the same or are rather similar when lipid environments are likely altered from one cell type to another. For example, when a voltage-gated potassium channel (Kv) is changed from a phospholipid-rich to phospholipid-free bilayer, we discovered that the voltage sensor domains of the channel were switched from one conformation (activated state) to the other (equivalent to a resting state) (Jiang, 2018, 2019; Zheng et al., 2012, 2016). The estimated energetic effects of a DOPC membrane containing 9.6 mol % cholesterol on a Kv channel may be as much as 15 kcal/mol, sufficient to change the states of its four voltage-sensor domains. Similarly, phosphatidylinositol-4,5-bisphosphate (PIP2), a lipid molecule being dynamically controlled in cell membranes, is known to participate in the controlled opening/closing of inward rectifier potassium channels, KCNQ1 Kv channels, TRPV4 channels, etc. (Harraz et al., 2018; Huang et al., 1998; Zaydman et al., 2013). Other membrane proteins, such as rhodopsin, 5-HT type 3 receptor (channel), type A GABA receptor and nicotinic acetylcholine receptor (nAchR), are all known to be sensitive to cholesterol in their response to ligands (Cascio, 2004; Mosqueira et al., 2018). These studies tend to highlight the importance of specific lipids in their binding pockets to the structural stability and proper functional property of these membrane proteins. But in a cell membrane, the local concentration of specific lipid molecules, such as cholesterol, glycolipids, and phospholipids, could be very high (100 mM or more) such that the specific high-affinity binding sites in a protein would be fully saturated. The lipids whose presence or absence modulates the functions of membrane proteins are usually those whose membrane content can be changed by enzymatic action, through lipid trafficking, or via domain switching within the same bilayer. From a thermodynamic sense, multiple low-affinity binding sites for a specific types of lipids around a protein could exert strong effects, sometimes far more significant than the lipids bound almost permanently at their high-affinity binding sites (Jiang, 2019).

A unit protein-centered microdomain (Figure 1.1) may contain only one copy of the target protein and dozens to hundreds of lipid molecules. Such a system is resistant to all available biochemical tools and below the sensitivity of almost all other techniques that fail to detect individual lipid molecules. This scenario has not changed much, even though SIMS (secondary ion mass spectrometry) imaging (Klitzing et al., 2013) is reaching tens of nm in spatial resolution and can detect sporadically liberated lipids from the membranes. New techniques with sufficient resolutions to detect individual lipid molecules in real time should be developed in order to study individual protein-centered domains in membranes.

1.3   TOPICS IN THE FOLLOWING CHAPTERS

In a more general sense, we would like to hypothesize that membrane proteins like Kv channels that are present in every live cell have been co-evolved with the lipid membranes whereby they play their physiological roles by adapting to local lipid environments. Geometrically, the lipid effects on membrane proteins may come from local, direct lipid-protein interactions or indirect changes in lipid bilayers such as surface tension, membrane fluidity, lipid phase changes, etc. The cell membranes may be geometrically complicated, especially in neuronal cells with arborized dendritic processes, and such geometric differences may exert the structural forces contributed by both lipid molecules and proteins in membranes. We therefore would generally expect location-variation of biomembranes from one site in a cell membrane to another might cause functional variations of individual membrane proteins, which could account for mode switches in the electrical activities of a single channel in the same patch over a long period of time. The same membrane proteins may be modulated differently among different sites in the same cell.

The geometrical factors and the local lipid molecules together manifest the interplay of direct and indirect interactions between lipids and membrane proteins. In order to experimentally study these interactions, we will need to understand the distribution of different types of lipid molecules around individual membrane proteins, and determine how the lipids and proteins interact with cytoskeleton network inside cells and extracellular matrix outside. So far, our analytical tools are not sufficient to fulfill this basic requirement for quantitative investigations of protein-lipid interactions.

To help overcome such technical limitations, I thought that it is important to look into the latest new technologies that are useful for studying biomembranes and lipid-protein interactions. In this book, new techniques and new case studies will be introduced by scientists working in the relevant areas. It will focus on different techniques, and a forthcoming book, *Lipid Determinants of Membrane-Protein Interactions*, will focus on the lipid effects on specific membrane proteins. This book contains seven topics. Chapters 2 and 3 will focus on two different high-sensitivity methods to measure lipids in cell membranes—lipidomic analysis, and surface-enhanced Raman spectroscopy (SERS) imaging. As you will discover, these methods still have a distance from the ultimate resolution of individual lipids. Chapters 4 and 5 will discuss two imaging techniques for biomembranes—scanning angle interference microscopy (SAIM), and high-speed atomic force microscopy (HS-AFM). Both techniques are suitable for looking into microdomains in cell membranes. Chapter 6 will introduce a set of *in vitro* reconstituted membrane systems that have been used for analyzing the activities of membrane proteins in membranes of varying lipid compositions. In Chapter 7, Dr. Dopico et al. will discuss two different lipid effects—direct (local) and indirect effects—on large-conductance Ca^{2+}-activated potassium (BK) channels. The last chapter by Dr. Wang will describe in depth the use of single particle cryo-electron microscopy (cryo-EM) to study membrane proteins in small vesicles, which is named restrained spherical construction (RSC) and allows the precise control of lipid composition.

ACKNOWLEDGMENTS

I would like to thank Dr. Michael X. Zhu at University of Texas Health Science Center for his suggestion of preparing the two volumes in order to highlight our keen interests in understanding the structural and functional effects of lipid-protein interactions. The research programs in my laboratory have been supported by National Institutes of Health (R21GM131231, R01GM111367, R01GM093271 & R01GM088745), AHA (12IRG9400019), CF Foundation (JIANG15G0), Welch Foundation (I-1684) and

Cancer Prevention and Research Institute of Texas (RP120474), National Science Foundation-founded Center for Big Learning (NSF-CBL), Southeast Consortium for Microscopy of Macromolecular Machines (U24GM116788 with Dr. Ken Taylor as PI), Southeast Center of Integrated Metabolomics (SECIM), and funds intramurally from UT Southwestern Medical Center and University of Florida.

REFERENCES

Bentley, E.P., Frey, B.B., and Deniz, A.A. (2019). Physical chemistry of cellular liquid-phase separation. *Chemistry 25*, 5600–5610.

Branton, D. (1966). Fracture faces of frozen membranes. *Proceedings of the National Academy of Sciences of the United States of America 55*, 1048–1056.

Callenberg, K.M., Latorraca, N.R., and Grabe, M. (2012). Membrane bending is critical for the stability of voltage sensor segments in the membrane. *The Journal of General Physiology 140*, 55–68.

Cascio, M. (2004). Structure and function of the glycine receptor and related nicotinicoid receptors. *The Journal of Biological Chemistry 279*, 19383–19386.

Danielli, J.F., and Davson, H. (1935). A contribution to the theory of permeability of thin films. *Journal of Cellular and Comparative Physiology 5*, 14.

Gomes, E., and Shorter, J. (2019). The molecular language of membraneless organelles. *The Journal of Biological Chemistry 294*, 7115–7127.

Harraz, O.F., Longden, T.A., Hill-Eubanks, D., and Nelson, M.T. (2018). PIP2 depletion promotes TRPV4 channel activity in mouse brain capillary endothelial cells. *eLife 7*, e38689.

Heimburg, T. (2007). *Thermal Biophysics of Membranes*, Vol. 1 (Berlin, Germany: Wiley-VCH Verlag GmbH & Co. KGaA).

Huang, C.L., Feng, S., and Hilgemann, D.W. (1998). Direct activation of inward rectifier potassium channels by PIP2 and its stabilization by Gbetagamma. *Nature 391*, 803–806.

Jacobson, K., Mouritsen, O.G., and Anderson, R.G. (2007). Lipid rafts: At a crossroad between cell biology and physics. *Nature Cell Biology 9*, 7–14.

Jiang, Q.X. (2018). Lipid-dependent gating of ion channels. In *Protein-lipid Interactions: Perspectives, Techniques and Challenges*, A. Catala, ed. (New York: Nova Science Publishers), p. 196.

Jiang, Q.X. (2019). Cholesterol-dependent gating effects on ion channels. *Advances in Experimental Medicine and Biology 1115*, 167–190.

Klitzing, H.A., Weber, P.K., and Kraft, M.L. (2013). Secondary ion mass spectrometry imaging of biological membranes at high spatial resolution. *Methods in Molecular Biology 950*, 483–501.

Mosqueira, A., Camino, P.A., and Barrantes, F.J. (2018). Cholesterol modulates acetylcholine receptor diffusion by tuning confinement sojourns and nanocluster stability. *Scientific Reports 8*, 11974.

Mouritsen, O.G., and Bloom, M. (1984). Mattress model of lipid-protein interactions in membranes. *Biophysical Journal 46*, 141–153.

Nicolson, G.L. (2014). The Fluid-Mosaic Model of membrane structure: Still relevant to understanding the structure, function and dynamics of biological membranes after more than 40 years. *Biochimica et Biophysica Acta 1838*, 1451–1466.

Robertson, J.D. (1967). Origin of the unit membrane concept. *Protoplasma 63*, 218–245.

Singer, S.J., and Nicolson, G.L. (1972). The fluid mosaic model of the structure of cell membranes. *Science 175*, 720–731.

Stoeckenius, W. (1962a). Some electron microscopical observations on liquid-crystalline phases in lipid-water systems. *The Journal of Cell Biology 12*, 221–229.

Stoeckenius, W. (1962b). Structure of the plasma membrane: An electron-microscope study. *Circulation 26*, 1066–1069.

Stoeckenius, W., and Engelman, D.M. (1969). Current models for the structure of biological membranes. *The Journal of Cell Biology 42*, 613–646.

Wee, C.L., Ulmschneider, M.B., and Sansom, M.S. (2010). Membrane/toxin interaction energetics via serial multiscale molecular dynamics simulations. *Journal of Chemical Theory and Computation 6*, 966–976.

Zaydman, M.A., Silva, J.R., Delaloye, K., Li, Y., Liang, H., Larsson, H.P., Shi, J., and Cui, J. (2013). Kv7.1 ion channels require a lipid to couple voltage sensing to pore opening. *Proceedings of the National Academy of Sciences of the United States of America 110*, 13180–13185.

Zheng, H., Lee, S., Llaguno, M.C., and Jiang, Q.X. (2016). bSUM: A bead-supported unilamellar membrane system facilitating unidirectional insertion of membrane proteins into giant vesicles. *The Journal of General Physiology 147*, 77–93.

Zheng, H., Liu, W., Anderson, L.Y., and Jiang, Q.X. (2012). Lipid-dependent gating of a voltage-gated potassium channel. *Nature Communications 2*, 250.

2 Lipidomics in Human Cancer and Malnutrition

Iqbal Mahmud and Timothy J. Garrett

CONTENTS

2.1 INTRODUCTION

The entire spectrum of lipid molecular species in any biological system, tissue, cell or fluid is called the lipidome, and the profiling/mapping of the lipidome is called lipidomics.[1,2] Lipidomics not only involves the full characterization of lipid molecular species, but also explains biological roles with respect to expression/regulation of genes/proteins involved in lipid metabolism and function.[3] Lipidomics can be either untargeted, in which case global lipid profiling is usually performed on complex biological mixtures, or targeted, in which lipids of interest are already known and the instrument is set to analyze only those of interest.[4,5] Lipids are a collection of an extremely heterogeneous group of molecules and have been divided into eight categories (fatty acyls, glycerolipids, glycerophospholipids, sphingolipids, sterol lipids, prenol lipids, saccharolipids, and polyketides) where ketoacyl or isoprene subunits are the two common building blocks[6-9] (Figure 2.1).

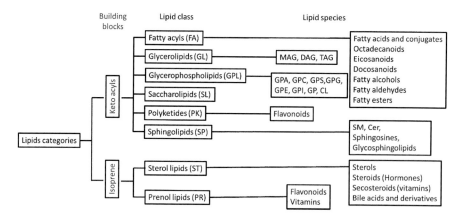

FIGURE 2.1 Major categories of lipids. Keto acyl and isoprene are two major building blocks of lipids where fatty acyls (FA), glycerolipids (GL), glycerophospholipids (GPL), saccharolipids (SL), polyketides (PK), and sphingolipids (SP) were derived from the condensation of keto acyl subunits, and sterol and prenol lipids from the condensation of isoprene subunits. Each lipid category contains diverse heterogeneous lipid species. MAG, monoacylglycerols; DAG, diacylglycerols; TAG, triacylglycerols; PA, phosphatidic acids; PC, Phosphatidylcholines; PS, Phosphatidylserines; PG, Phosphatidylglycerols; PE, Phosphatidylethanolamines; PI, Phosphatidylinositols; CL, Cardiolipins; SM, sphingomyolins; Cer, ceramides.

Lipids are vital components of cell membrane and play many essential roles in cellular functions including cellular barriers, membrane biogenesis, homeostasis, energy storage, and signaling pathways.[4,5,10–13] Emerging research has revealed critical roles of these lipids associated with cellular processes in human health and diseases including cancer,[14–16] lipid metabolism disorders,[17] malnutrition,[18,19] lipid storage diseases,[20] microbiota dysbiosis,[21,22] obesity,[23,24] heart attack,[25,26] brain stroke,[27] diabetes,[28] atherosclerosis,[26] and lipid pneumonia[29,30] (Figure 2.2).

There are multiples underlying pathways that lipids follow to disease onset (Figure 2.3). However, these liabilities of lipids in diverse human diseases are not clearly elucidated due to the extreme heterogeneity of lipid molecular structures and function. Lipidomics-based technologies could shed light on better identification of lipid structure and function and ultimately aid in developing new strategies to combat these lipid-associated health problems.[1] Recently, High-Resolution Mass Spectrometry (HRMS) and its aligned technologies have made significant improvement in lipidomics with increased coverage, structural elucidation of double-bond positional or *sn* positional isomers, high-throughput detection, identification, and accurate unknown discovery.[31–35] In this chapter, we explain the current knowledge of HRMS in lipid research with a focused discussion on HRMS-based lipidomics in human malnutrition and cancer, towards diagnosis and prediction of potential therapies.

FIGURE 2.2 Lipids in human health and diseases. The illustration encompasses the nine major lipid-associated human health problems. The past decade has observed remarkable development towards the understanding of lipid biology in human health and diseases through analytical technologies. NFLD, non-alcoholic fatty liver disease; FLD, Fatty liver disease.

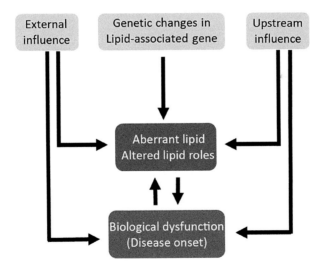

FIGURE 2.3 Mechanism of action in lipid metabolism and disease onset. Genetic alterations such as loss or gain of function mutation in lipid-associated genes or proteins involved in aberrant lipid metabolism. Upstream gene regulatory machineries involved in lipid-related gene transcription are associated with reprogramed lipid metabolism. External factors such as environmental factors, microbiota, and life style are also associated with aberrant lipid metabolism. Altered lipid metabolism then pursues disease onset. Upstream or external factors are directly related to biological dysfunction and then follow aberrant lipid metabolism.

2.2 HIGH-RESOLUTION MASS SPECTROMETRY (HRMS) AND ITS ALIGNED TOOLS IN LIPIDOMICS

HRMS is highly selective since it measures the exact mass of a lipid species. HRMS in conjunction with different separation techniques contributes to lipid data generation, processing and analysis, and leads to a meaningful connection to biological phenomena. In lipidomics, there are three complementary approaches in mass spectrometry-based lipid analysis such as direct infusion or shotgun lipidomics, liquid chromatography coupled mass spectrometry-based lipidomics (LC-MS), and MS imaging. These approaches and associated data analysis tools are summarized in Table 2.1.

2.2.1 SHOTGUN LIPIDOMICS

Tandem MS is a commonly used shotgun lipidomics technique which is also known as MS/MS or MS^n where samples are ionized by electrospray ionization as a mixture and several precursor ions via precursor ion scans (PIS) or selected product ion scans are utilized to separate out the various lipid species without then use of chromatographic separation.[5] Shotgun lipidomics analysis first collects a full-scan mass spectrum that displays molecular ions of individual lipid species of a class and then uses tandem MS spectra of corresponding samples to provide detailed structural and quantitative analysis without the time constraints encountered by LC-MS analysis. Tandem MS offers PIS of particular fragment ions, neutral loss scanning (NLS) of specific neutral loss fragments, and product ion scanning of molecular ions of interest; all these approaches facilitate the high-throughput analysis of global lipidome from complex biological mixtures.[36] A detailed review of this approach was recently published.[37]

2.2.2 LIQUID CHROMATOGRAPHY COUPLED HIGH-RESOLUTION MASS SPECTROMETRY (LC-HRMS)

Mass spectrometry-based lipidomics research has made tremendous advancement through coupling with liquid chromatographic separations, especially with the use of electrospray ionization (ESI) coupled to high-resolution mass spectrometry (LC-HRMS).[5,38] LC-HRMS is often considered the gold standard analytical technique in profiling lipidomic studies. LC-HRMS lipidomics identifies lipid species by both chromatographic separation and mass spectral data. LC adds retention time, which is a valuable for the enrichment of low abundance species and separation of potential isomers that enables identification by accurate mass and MS/MS.[39] As an example, LC-HRMS can be efficiently applied to identify expression analysis of diverse lipids including glycerophospholipid, and subsequent data-dependent HRMS technique (ddMS/MS) can be utilized for fragmentation and structural elucidation of lipid species with *in silico* based libraries harnessing the common fragmentation pathways of each lipid class. An example identification of PC 16:0_16:1 is displayed in Figure 2.4 with the analysis of lipids in cancer. Chromatographic separation (Figure 2.4, top) provides for the separation of lipid species and then performing HRMS/MS enables identification of the lipid based on known fragmentation pathways (Figure 2.4, bottom).

TABLE 2.1

Examples of Mass Spectrometry-Based Lipidomics Techniques and Analysis Tools

MS-Based Lipidomics Data Acquisition Techniques

MS Technique	MS Types	Ionization	Comments
Direct or shotgun MS	Tandem MS	ESI	Analysis without the time constraints
	MALDI-MS	MALDI	Provide tremendous sensitivity and selectivity
	Ion-mobility MS		Targets wide and complex array of lipid species
Chromatography based MS	LC-MS	ESI	Add retention time in MS data separation
	LC-IM-MS	APCI MALDI	Add additional dimension of separation
Imaging MS	MALDI-MSI	MALDI	Imaging lipids under laser condition
	DESI-MSI	DESI	Imaging lipids at ambient condition

Raw Data Processing Tools

Name	Use	Availability
msConvert	Read raw data files into open data formats	http://proteowizard.sourceforge.net/tools.shtml
RawConverter	It also extracts raw MS data into open data formats	http://fields.scripps.edu/rawconv/

Lipidomics Data Analysis Tools

Name	Use	Availability
IE-Omics	Expands lipid coverage	http://secim.ufl.edu/secim-tools/ie-omics/
LipidMatch	Lipid identification and analysis	http://secim.ufl.edu/secim-tools/lipidmatch/
LipidMatch Flow	Provide entire lipidomics data-processing workflow	http://secim.ufl.edu/secim-tools/lipidmatchflow/
LipidMatch Normalizer	Performs lipid class specific relative quantification of lipid	http://secim.ufl.edu/secim-tools/lipidmatch-normalizer/
Lipid Pioneer	Provides exact mass & molecular formula of lipid species	http://secim.ufl.edu/secim-tools/lipidpioneer/
LipidQC	Rapidly compare lipid concentration measurements	http://secim.ufl.edu/secim-tools/lipidqc/
MZMine	Processing and analysis	http://mzmine.github.io/
XCMS	Processing and analysis	https://xcmsonline.scripps.edu

Biological Interpretation Analysis Tools

Name	Use	Availability
MetaboAnalyst	Pathway and biomarker analysis	https://www.metaboanalyst.ca/MetaboAnalyst/faces/home.xhtml
NetworkAnalyst	Lipid network analysis	https://www.networkanalyst.ca/
OmicsNet	Gene-lipid integrated pathway analysis	https://www.omicsnet.ca/

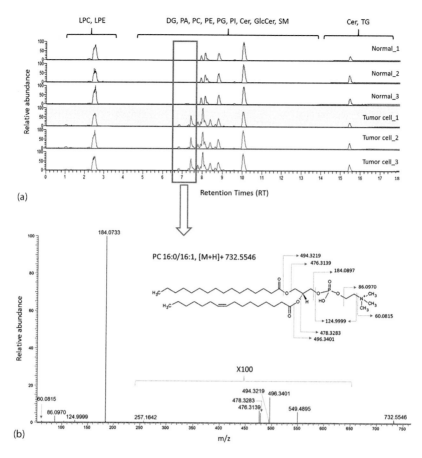

FIGURE 2.4 Expression and elucidation of lipid structure analysis by LC-HRMS/MS. (a) Comparative overlays of three replications of each oral normal and oral tumor cell chromatograms are shown. Rectangle box following arrow indicates significantly increased peaks in tumor cells compare to normal. (b) Using ddMS2, the peak (7.5 min) in green box of Figure 2.4a is identified as PC (16:0/16:1), where its complete structural elucidation including head group, choline head group, phosphate moiety, glycerol back bone, and fatty acyl tails is shown as arrows. Precursor *m/z* value was 732.5545.

This lipid is correctly identified by the presence of *m/z* 184.0733, which is representative of the head group for a phosphatidylcholine and the fatty acyl chains are identified by the less adundant ions of *m/z* 478.3283 and 496.3401 for 16:1 and *m/z* 476.3139 and 494.3216 for 16:0 (inset of Figure 2.4 shows the fragmentation pathways).

2.2.3 MASS SPECTROMETRY IMAGING (MSI)-BASED LIPIDOMICS

MSI is an emerging tool in lipidomics research, which allows untargeted analysis and structural characterization of lipids directly from a tissue section. There are several MSI analysis approaches, but the two most common are Desorption Electrospray

Ionization Mass Spectrometry Imaging (DESI-MSI) and Matrix-Assisted Laser Desorption/Ionization Mass Spectrometry Imaging (MALDI-MSI) where DESI-MSI occurs under ambient conditions and MALDI-MSI most often is accomplished under vacuum although atmospheric pressure MALDI is still performed.[40–43]

In general, both ESI and MALDI are the most common ionization mechanisms for both shotgun and LC-MS approaches. Moreover, shotgun and LC-MS techniques are both compatible with ion mobility MS.[44] Ion-mobility MS (IMS) provides several benefits over shotgun or LC-MS alone, which includes separation of isobaric and isomeric lipids by gas-phase structure which is a completely orthogonal approach and thus can provide simplification of complex extracts analysis through addition/reduction of the complexity of lipid mixtures, and increasing selectivity by providing separation of endogenous matrix interferences.[5,44]

2.2.4 RECENT ADVANCEMENT IN LIPIDOMICS DATA PROCESSING AND ANALYSIS TOOLS

Data processing tools typically perform data quality control and analysis, which includes mass detection, chromatogram building and deconvolution, noise filtering, baseline correction, peak detection, deisotoping, alignment, gap-filling, accurate mass calculation, identification of lipid species, data normalization, visualization, and biological interpretation.

Previously, our group and others have developed Ultra-High Performance Liquid Chromatography coupled with High-Resolution Mass Spectrometry (UHPLC-HRMS)-based untargeted and targeted lipidome profiling protocols, analytical technologies, and its associated bioinformatics tools.[31–35,45] As an example, LipidMatch, an interactive lipid data analysis software, is utilized to annotate lipids detected from LC-HRMS/MS data-dependent and data-independent (DIA) experiments using various vendor formats.

LipidMatch can also be employed for direct infusion HRMS/MS and imaging experiments.[32] LipidMatch Flow, a user-friendly software covering the entire lipidomics data-processing workflow for LC-HRMS/MS data analysis, incorporates MSConvert for file conversion, MZmine for peak picking optimized for each vendor, LipidMatch for identification, and combines adducts and polarities into a single file, which can be uploaded into metaboanalyst for statistics.[31] LipidMatch Normalizer (LMN), a lipid data normalization tool, can be used to perform relative quantification for lipidomics using class specific lipid standards and LC-MS data. The user provides a feature table containing all the annotated lipids and their respective m/z values, retention time, and intensities across samples. LipidQC, a Microsoft Excel-based (MSXL) visualization tool, provides a means—independent of sample preparation methods, MS instruments, and lipid-adduct—to rapidly compare lipid concentration measurements (nmol/mL) with the available mean value estimates for four NIST Standard Reference Materials.[34]

LipidPioneer, an interactive template, can be used to generate exact masses and molecular formulas of lipid species that may be encountered in the mass spectrometric analysis of lipid profiles. Over 60 lipid classes are present in the LipidPioneer template, and include several unique lipid species, such as ether-linked lipids and lipid oxidation products. In the template, users can add any fatty acyl constituents without limitation in the number of carbons or degrees of unsaturation.

2.3   LC-HRMS-BASED LIPIDOMICS IN CHILDHOOD MALNUTRITION

Childhood malnutrition is a serious health problem for children in developing countries.[46] Undernutrition, a major form of malnutrition, leads to stunted growth of over 250 million (43%) children under 5 years of age worldwide.[47,48] Stunted growth is associated with numerous child health problems including impaired cognitive and motor development, neurodevelopment, morbidity, and mortality.[19,48] LC-MS-based targeted lipidomics has been conducted in childhood undernutrition studies and identified dysregulated lipids and their roles in children's stunted growth and disease pathogenesis[18,19,49] (Figure 2.5). Using LC-MS techniques, Semba et al. first identified low ω-3 and ω-6 long-chain polyunsaturated fatty acids (PUFAs), which are essential for growth and development and low carnitine, which is essential for β-oxidation of fatty acids, in stunted children from Malawi compare to normal.[18] Another study identified that both sphingomyelins (SM) and phosphatidylcholines (PC) were associated with improved growth/developmental outcomes where elevated levels of PC were observed with higher neurocognitive scores.[19,50] That study first demonstrated that separate metabolic pathways might relate to stunting and neurocognition. Lower levels of sulfated neurosteroids in serum samples of stunted children were observed and indicated a potential association between stunting and brain development.[18] Gut microbiota dysbiosis is a common problem in developing countries with malnourished children.[51] LC-MS/MS analysis identified reduced glycerolipid and glycerophospholipids production capabilities in children from two low-income countries compared with high-income countries and suggested future efforts towards further characterization of gut microbial metabolic irregularities and their contribution to malnutrition.[52] Overall, LC-MS-based lipidomics appears to

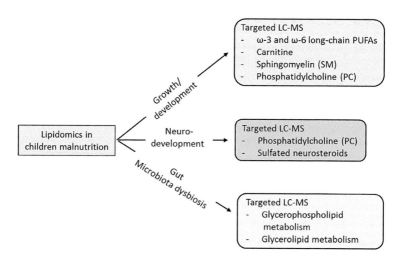

FIGURE 2.5   LC-HRMS-based lipidomics in childhood malnutrition and pathogenesis. Three important directions such as children's growth, neurodevelopment, and gut microbiota dysbiosis that LC-MS-based targeted lipidomics demonstrated the potential roles of lipids in childhood malnutrition.

be emerging in understanding the roles of lipids in biology and the pathogenesis of child malnutrition. However, our understanding of lipidomics in child malnutrition is currently limited to predetermined targeted lipid molecules. Therefore, it is highly important to study the global lipid profile in an untargeted manner, which may help us better understand the dysfunction of diverse lipids and their associated metabolic networks in child malnutrition.

2.4 LC-HRMS-BASED LIPIDOMICS IN CANCER

Of the ten leading causes of death, cancer is the deadliest form of public health problem in the United States and worldwide.[53] In 2018, 1,735,350 new cancer cases and 609,640 cancer deaths were projected to occur in the United States.[53] Currently, there is no efficient and safe treatment option for the management of cancer. Lipids, an important class of biomolecules, play critical roles during tumor initiation and disease progression.[15] They are involved in diverse oncogenic functions such as transformation of normal epithelial cells to tumorigenic cells, disruption of normal tissue architecture,[54] cell migration,[55] cell invasion,[56] etc., and act as homing signals for tumor cells, angiogenesis,[57] metastasis, and drug resistence.[15,58] Therefore, LC-HRMS-based lipidomic studies have gained significant attraction regarding the discovery of potential lipid biomarkers for early cancer screening, diagnosis, prognosis, and new treatments.

HRMS-based targeted or untargeted lipidomics have been applied in many cancers for understanding oncogenic roles of lipids or regulation of lipids by oncogenes in cancer. Using ultra-high pressure liquid chromatography coupled high-resolution mass spectrometry (UHPLC-HRMS), Mahmud et al. identified Death-Domain Associated protein (DAXX) as an oncogene and determined that it interacts with and co-activates sterol regulatory elements binding proteins (SREBPs) and drives *de novo* lipogenesis including fatty acids, glycerolipids, glycerophospholipids, and cholesterol biosynthesis in diverse cancers, such as triple negative breast cancer.[58,59] Moreover, UHPLC-HRMS-based lipidomics revealed, SR-4370, novel optimized analog of UF010, a potent inhibitor of class I histone deacetylases (HDACs 1-3), exhibited potent inhibition of lipid production in cells and its corresponding xenograft tumors of diverse cancer types.[60] Similarly, MYC oncogene and mutant p53 were found to elevate lipid production in diverse cancer types via SREBPs co-activation.[54,61]

Aberrant synthesis of lipids has been associated with tumor biology and multiple oncogenic responses[15] (Figure 2.6). For example, dysregulation of mevalonate, cholesterol, eicosanoids, and fatty acid synthesis pathways is associated with cellular transformation.[62–65] Disruption of normal tissue structures is a cancer hallmark. Lipid contents including mevalonate and cholesterol levels were related to a loss of normal tissue architecture.[54] Secondly, aberrant cancer cell proliferation requires the rapid synthesis of lipids for the generation of biological membranes and cell division. Various lipid molecules such as prostaglandins (PG) PGE2, PGE3, eicosanoids, phospholipids, lysophosphatidic acids (LPA), fatty acids, sphingosine-1-phosphate (S1P), cholesterol, 27-hydroxycholesterol promote rapid proliferation of multiple tumor cells.[66–71]

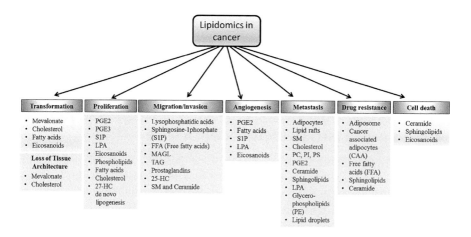

FIGURE 2.6 Role of lipids in tumor biology. Examples of lipids involved in diverse tumor biological processes including cellular transformation, cell proliferation, migration, invasion, angiogenesis, metastasis, drug resistance and cell death.

Thirdly, tumor cell migration or invasion into surrounding tissue and the vasculature is an initial step in progressive or metastatic nature of tumor biology. Protrusion activity of cell membrane and disruption of the extracellular matrix are critical for cell migration or invasion.[72] Lipids including LPA, SM, Ceramide, S1P, MAGL (monoacylglycerol), DAG (Diacylglycerol), TAG (triacylglycerol), prostaglandins, and 25-hydroxycholesterol promote both migration and invasion of diverse cancers.[15,72–75] Fourthly, induction of angiogenesis is an essential mechanism for tumor growth and survival.[76] PGE2 promotes angiogenesis by stimulating proliferation and tube formation in endothelial cells.[57] Other lipids such as S1P, fatty acids, LPA, and eicosanoids were also found to promote angiogenesis in several cancers.[65,76,77] Moreover, metastasis is the most deadly aspect of cancer and is responsible for more than 90% of all cancer-related deaths.[78,79] A large number of lipids or lipid-bodies such as adipocytes (containing TAG, DAG, phospholipids, and cholesterol ester), lipids rafts, sphingolipids, cholesterol, glycerophospholipids (PC, PE, PI, PS), LPA, prostaglandins, and fatty acids significantly contribute to metastatic cancer.[15,57,80–82] Cancer cells store excess lipids in the form of lipid droplets or adiposome which supply energy to power tumor expansion and metastasis.[83,84] Cancer-associated adipocytes (CAA) or lipid droplets, sphingolipids, and free fatty acids contribute to the resistance of anti-cancer drugs.[69,85] Furthermore, lipid molecules not only are involved in different oncogenic pathways, but also are related to tumor cell death. For example, ceramide and sphingolipids potentiate growth inhibitory signaling mechanisms that drive apoptotic cell death.[69]

2.5 SUMMARY AND FUTURE PERSPECTIVES IN LIPIDOMICS

In summary, although lipidomics in child malnutrition and cancer are at an early stage compared to other omics, they have made a tremendous contribution to elucidating disease pathogenesis and providing diagnostic potential. In child malnutrition, targeted lipidomics has started making a significant contribution via understanding roles of lipids in children's growth, gut microbiota dysbiosis, and brain development. Future untargeted lipidomics studies in child malnutrition could provide better insights on global functions of lipids in children with malnourishment.[86] In cancer, both targeted and untargeted lipidomics have been conducted and have made significant advancement in understanding roles of lipids in tumor biology and cancer therapeutics. One of the major drawback in lipidomics research is to identify unknowns, the metabolic dark matter. Based on current lipidomics practice, over 80% of metabolites or lipids are unknown.[87] Therefore, identification of metabolic dark matter in lipid research using the state-of-the-art analytical instrumentation will reveal new potential in next-generation metabolism research.

ACKNOWLEDGMENT

Research reported in this chapter was supported by the University of Florida Clinical and Translational Science Institute, which is supported in part by the NIH National Center for Advancing Translational Sciences under award number UL1TR001427. The content is solely the responsibility of the authors and does not necessarily represent the official views of the National Institutes of Health.

REFERENCES

1. Spener, F., Lagarde, M., Géloên, A., Record, M. Editorial: What Is Lipidomics? *Eur. J. Lipid Sci. Technol.* **2003**, *105* (9), 481–482. https://doi.org/10.1002/ejlt.200390101.
2. Lagarde, M., Géloën, A., Record, M., Vance, D., Spener, F. Lipidomics Is Emerging. *Biochim. Biophys. Acta BBA—Mol. Cell Biol. Lipids* **2003**, *1634* (3), 61. https://doi.org/10.1016/j.bbalip.2003.11.002.
3. Dennis, E. A. Lipidomics Joins the Omics Evolution. *Proc. Natl. Acad. Sci. U. S. A.* **2009**, *106* (7), 2089–2090. https://doi.org/10.1073/pnas.0812636106.
4. Wenk, M. R. Lipidomics: New Tools and Applications. *Cell* **2010**, *143* (6), 888–895. https://doi.org/10.1016/j.cell.2010.11.033.
5. Yang, K., Han, X. Lipidomics: Techniques, Applications, and Outcomes Related to Biomedical Sciences. *Trends Biochem. Sci.* **2016**, *41* (11), 954–969. https://doi.org/10.1016/j.tibs.2016.08.010.
6. Fahy, E., Cotter, D., Sud, M., Subramaniam, S. Lipid Classification, Structures and Tools. *Biochim. Biophys. Acta* **2011**, *1811* (11), 637–647. https://doi.org/10.1016/j.bbalip.2011.06.009.
7. Sud, M., Fahy, E., Cotter, D., Brown, A., Dennis, E. A., Glass, C. K., Merrill, A. H., Murphy, R. C., Raetz, C. R. H., Russell, D. W. et al. LMSD: LIPID MAPS Structure Database. *Nucleic Acids Res.* **2007**, *35* (suppl_1), D527–D532. https://doi.org/10.1093/nar/gkl838.
8. Fahy, E., Subramaniam, S., Murphy, R. C., Nishijima, M., Raetz, C. R. H., Shimizu, T., Spener, F., van Meer, G., Wakelam, M. J. O., Dennis, E. A. Update of the LIPID MAPS Comprehensive Classification System for Lipids. *J. Lipid Res.* **2009**, *50* (Suppl), S9–S14. https://doi.org/10.1194/jlr. R800095-JLR200.
9. Fahy, E., Subramaniam, S., Brown, H. A., Glass, C. K., Merrill, A. H., Murphy, R. C., Raetz, C. R. H., Russell, D. W., Seyama, Y., Shaw, W. et al. A Comprehensive Classification System for Lipids. *J. Lipid Res.* **2005**, *46* (5), 839–861. https://doi.org/10.1194/jlr. E400004-JLR200.

10. Sampaio, J. L., Gerl, M. J., Klose, C., Ejsing, C. S., Beug, H., Simons, K., Shevchenko, A. Membrane Lipidome of an Epithelial Cell Line. *Proc. Natl. Acad. Sci. U. S. A.* **2011**, *108* (5), 1903–1907. https://doi.org/10.1073/pnas.1019267108.

11. van Meer, G., Voelker, D. R., Feigenson, G. W. Membrane Lipids: Where They Are and How They Behave. *Nat. Rev. Mol. Cell Bio.* **2008**, 9, 112–124.

12. Harayama, T., Riezman, H. Understanding the Diversity of Membrane Lipid Composition. *Nat. Rev. Mol. Cell Bio.* **2018**, 19, 281.

13. Storck, E. M., Özbalci, C., Eggert, U. S. Lipid Cell Biology: A Focus on Lipids in Cell Division. *Annu. Rev. Biochem.* **2018**, *87*, 839–869. https://doi.org/10.1146/annurev-biochem-062917-012448.

14. Perrotti, F., Rosa, C., Cicalini, I., Sacchetta, P., Del Boccio, P., Genovesi, D., Pieragostino, D. Advances in Lipidomics for Cancer Biomarkers Discovery. *Int. J. Mol. Sci.* **2016**, *17* (12). https://doi.org/10.3390/ijms17121992.

15. Baenke, F., Peck, B., Miess, H., Schulze, A. Hooked on Fat: The Role of Lipid Synthesis in Cancer Metabolism and Tumour Development. *Dis. Model Mech.* **2013**, *6* (6), 1353–1363.

16. Currie, E., Schulze, A., Zechner, R., Walther, T. C., Farese, R. V. Cellular Fatty Acid Metabolism and Cancer. *Cell Metab.* **2013**, *18* (2), 153–161. https://doi.org/10.1016/j.cmet.2013.05.017.

17. Amor, A. J., Perea, V. Dyslipidemia in Nonalcoholic Fatty Liver Disease. *Curr. Opin. Endocrinol. Diabetes Obes.* **2019**, *26* (2), 103–108. https://doi.org/10.1097/MED.0000000000000464.

18. Semba, R. D., Trehan, I., Li, X., Salem, N., Moaddel, R., Ordiz, M. I., Maleta, K. M., Kraemer, K., Manary, M. J. Low Serum ω-3 and ω-6 Polyunsaturated Fatty Acids and Other Metabolites Are Associated with Poor Linear Growth in Young Children from Rural Malawi. *Am. J. Clin. Nutr.* **2017**, *106* (6), 1490–1499. https://doi.org/10.3945/ajcn.117.164384.

19. Moreau, G. B., Ramakrishnan, G., Cook, H. L., Fox, T. E., Nayak, U., Ma, J. Z., Colgate, E. R., Kirkpatrick, B. D., Haque, R., Petri, W. A. Childhood Growth and Neurocognition Are Associated with Distinct Sets of Metabolites. *EBioMed.* **2019**. https://doi.org/10.1016/j.ebiom.2019.05.043.

20. Schulze, H., Sandhoff, K. Lysosomal Lipid Storage Diseases. *Cold Spring Harb. Perspect. Biol.* **2011**, *3* (6). https://doi.org/10.1101/cshperspect.a004804.

21. Ghazalpour, A., Cespedes, I., Bennett, B. J., Allayee, H. Expanding Role of Gut Microbiota in Lipid Metabolism. *Curr. Opin. Lipidol.* **2016**, *27* (2), 141–147. https://doi.org/10.1097/MOL.0000000000000278.

22. Kindt, A., Liebisch, G., Clavel, T., Haller, D., Hörmannsperger, G., Yoon, H., Kolmeder, D., Sigruener, A., Krautbauer, S., Seeliger, C. et al. The Gut Microbiota Promotes Hepatic Fatty Acid Desaturation and Elongation in Mice. *Nat. Commun.* **2018**, 9 (1), 1–15. https://doi.org/10.1038/s41467-018-05767-4.

23. González-Muniesa, P., Mártinez-González, M.-A., Hu, F. B., Després, J.-P., Matsuzawa, Y., Loos, R. J. F., Moreno, L. A., Bray, G. A., Martinez, J. A. Obesity. *Nat. Rev. Dis. Primer* **2017**, *3* (1), 1–18. https://doi.org/10.1038/nrdp.2017.34.

24. Klop, B., Elte, J. W. F., Castro Cabezas, M. Dyslipidemia in Obesity: Mechanisms and Potential Targets. *Nutrients* **2013**, *5* (4), 1218–1240. https://doi.org/10.3390/nu5041218.

25. Arsenault, B. J., Boekholdt, S. M., Kastelein, J. J. P. Lipid Parameters for Measuring Risk of Cardiovascular Disease. *Nat. Rev. Cardiol.* **2011**, *8* (4), 197–206. https://doi.org/10.1038/nrcardio.2010.223.

26. Linton, M. F., Yancey, P. G., Davies, S. S., Jerome, W. G., Linton, E. F., Song, W. L., Doran, A. C., Vickers, K. C. The Role of Lipids and Lipoproteins in Atherosclerosis. In *Endotext*, Feingold, K. R., Anawalt, B., Boyce, A., Chrousos, G., Dungan, K., Grossman, A., Hershman, J. M., Kaltsas, G., Koch, C., Kopp, P. et al., Eds.; MDText. com: South Dartmouth, MA, 2000.

27. Hindy, G., Engström, G., Larsson, S. C., Traylor, M., Markus, H. S., Melander, O., Orho-Melander, M. Role of Blood Lipids in the Development of Ischemic Stroke and Its Subtypes. *Stroke* **2018**, *49* (4), 820–827. https://doi.org/10.1161/STROKEAHA.117.019653.

28. Mooradian, A. D. Dyslipidemia in Type 2 Diabetes Mellitus. *Nat. Clin. Pract. Endocrinol. Metab.* **2009**, *5* (3), 150–159. https://doi.org/10.1038/ncpendmet1066.

29. Marangu, D., Gray, D., Vanker, A., Zampoli, M. Exogenous Lipoid Pneumonia in Children: A Systematic Review. *Paediatr. Respir. Rev.* **2019**. https://doi.org/10.1016/j.prrv.2019.01.001.

30. Sung, S., Tazelaar, H. D., Crapanzano, J. P., Nassar, A., Saqi, A. Adult Exogenous Lipoid Pneumonia: A Rare and Underrecognized Entity in Cytology—A Case Series. *CytoJournal* **2018**, *15*, 17. https://doi.org/10.4103/cytojournal.cytojournal_29_17.

31. Koelmel, J. P., Kroeger, N. M., Gill, E. L., Ulmer, C. Z., Bowden, J. A., Patterson, R. E., Yost, R. A., Garrett, T. J. Expanding Lipidome Coverage Using LC-MS/MS Data-Dependent Acquisition with Automated Exclusion List Generation. *J. Am. Soc. Mass Spectrom.* **2017**, *28* (5), 908–917. https://doi.org/10.1007/s13361-017-1608-0.

32. Koelmel, J. P., Kroeger, N. M., Ulmer, C. Z., Bowden, J. A., Patterson, R. E., Cochran, J. A., Beecher, C. W. W., Garrett, T. J., Yost, R. A. LipidMatch: An Automated Workflow for Rule-Based Lipid Identification Using Untargeted High-Resolution Tandem Mass Spectrometry Data. *BMC Bioinformatics* **2017**, *18* (1), 331. https://doi.org/10.1186/s12859-017-1744-3.

33. Koelmel, J. P., Cochran, J. A., Ulmer, C. Z., Levy, A. J., Patterson, R. E., Olsen, B. C., Yost, R. A., Bowden, J. A., Garrett, T. J. Software Tool for Internal Standard Based Normalization of Lipids, and Effect of Data-Processing Strategies on Resulting Values. *BMC Bioinformatics* **2019**, *20* (1), 217. https://doi.org/10.1186/s12859-019-2803-8.

34. Ulmer, C. Z., Ragland, J. M., Koelmel, J. P., Heckert, A., Jones, C. M., Garrett, T. J., Yost, R. A., Bowden, J. A. LipidQC: Method Validation Tool for Visual Comparison to SRM 1950 Using NIST Interlaboratory Comparison Exercise Lipid Consensus Mean Estimate Values. *Anal. Chem.* **2017**, *89* (24), 13069–13073. https://doi.org/10.1021/acs.analchem.7b04042.

35. Ulmer, C. Z., Patterson, R. E., Koelmel, J. P., Garrett, T. J., Yost, R. A. A Robust Lipidomics Workflow for Mammalian Cells, Plasma, and Tissue Using Liquid-Chromatography High-Resolution Tandem Mass Spectrometry. In *Lipidomics: Methods and Protocols*, Bhattacharya, S. K., Ed.; *Methods in Molecular Biology*; Springer: New York, 2017, pp. 91–106. https://doi.org/10.1007/978-1-4939-6996-8_10.

36. Yang, K., Han, X. Accurate Quantification of Lipid Species by Electrospray Ionization Mass Spectrometry—Meets a Key Challenge in Lipidomics. *Metabolites* **2011**, *1* (1), 21–40. https://doi.org/10.3390/metabo1010021.

37. Hsu, F.-F. Mass Spectrometry-Based Shotgun Lipidomics: A Critical Review from the Technical Point of View. *Anal. Bioanal. Chem.* **2018**, *410* (25), 6387–6409. https://doi.org/10.1007/s00216-018-1252-y.

38. Wang, C., Wang, M., Han, X. Applications of Mass Spectrometry for Cellular Lipid Analysis. *Mol. Biosyst.* **2015**, *11* (3), 698–713. https://doi.org/10.1039/c4mb00586d.

39. Cajka, T., Fiehn, O. Comprehensive Analysis of Lipids in Biological Systems by Liquid Chromatography-Mass Spectrometry. *TrAC Trends Anal. Chem.* **2014**, *61*, 192–206. https://doi.org/10.1016/j.trac.2014.04.017.

40. Murphy, R. C., Hankin, J. A., Barkley, R. M. Imaging of Lipid Species by MALDI Mass Spectrometry. *J. Lipid Res.* **2009**, *50* (*Suppl*), S317–S322. https://doi.org/10.1194/jlr.R800051-JLR200.

41. Eberlin, L. S., Ferreira, C. R., Dill, A. L., Ifa, D. R., Cooks, R. G. Desorption Electrospray Ionization Mass Spectrometry for Lipid Characterization and Biological Tissue Imaging. *Biochim. Biophys. Acta* **2011**, *1811* (11), 946–960. https://doi.org/10.1016/j.bbalip.2011.05.006.

42. Garrett, T. J., Dawson, W. W. Lipid Geographical Analysis of the Primate Macula by Imaging Mass Spectrometry. *Methods Mol. Biol. Clifton NJ* **2009**, *579*, 247–260. https://doi.org/10.1007/978-1-60761-322-0_12.

43. Garrett, T. J., Menger, R. F., Dawson, W. W., Yost, R. A. Lipid Analysis of Flat-Mounted Eye Tissue by Imaging Mass Spectrometry with Identification of Contaminants in Preservation. *Anal. Bioanal. Chem.* **2011**, *401* (1), 103–113. https://doi.org/10.1007/s00216-011-5044-x.

44. Paglia, G., Kliman, M., Claude, E., Geromanos, S., Astarita, G. Applications of Ion-Mobility Mass Spectrometry for Lipid Analysis. *Anal. Bioanal. Chem.* **2015**, *407* (17), 4995–5007. https://doi.org/10.1007/s00216-015-8664-8.

45. Koelmel, J. P., Ulmer, C. Z., Fogelson, S., Jones, C. M., Botha, H., Bangma, J. T., Guillette, T. C., Luus-Powell, W. J., Sara, J. R., Smit, W. J. et al. Lipidomics for Wildlife Disease Etiology and Biomarker Discovery: A Case Study of Pansteatitis Outbreak in South Africa. *Metabolomics* **2019**, *15* (3), 38. https://doi.org/10.1007/s11306-019-1490-9.

46. Bhutta, Z. A., Berkley, J. A., Bandsma, R. H. J., Kerac, M., Trehan, I., Briend, A. Severe Childhood Malnutrition. *Nat. Rev. Dis. Primer* **2017**, *3*, 17067. https://doi.org/10.1038/nrdp.2017.67.

47. Black, R. E., Allen, L. H., Bhutta, Z. A., Caulfield, L. E., de Onis, M., Ezzati, M., Mathers, C., Rivera, J. Maternal and Child Undernutrition: Global and Regional Exposures and Health Consequences. *Lancet* **2008**, *371* (9608), 243–260. https://doi.org/10.1016/S0140-6736(07)61690-0.

48. Black, M. M., Walker, S. P., Fernald, L. C. H., Andersen, C. T., DiGirolamo, A. M., Lu, C., McCoy, D. C., Fink, G., Shawar, Y. R., Shiffman, J. et al. Early Childhood Development Coming of Age: Science through the Life Course. *Lancet* **2017**, *389* (10064), 77–90. https://doi.org/10.1016/S0140-6736(16)31389-7.

49. Semba, R. D., Shardell, M., Trehan, I., Moaddel, R., Maleta, K. M., Ordiz, M. I., Kraemer, K., Khadeer, M., Ferrucci, L., Manary, M. J. Metabolic Alterations in Children with Environmental Enteric Dysfunction. *Sci. Rep.* **2016**, *6*, 28009. https://doi.org/10.1038/srep28009.

50. van den Heuvel, M. Metabolomics, Stunting and Neurodevelopment. *EBioMedicine* **2019**, *44*, 10–11. https://doi.org/10.1016/j.ebiom.2019.05.067.

51. Subramanian, S., Huq, S., Yatsunenko, T., Haque, R., Mahfuz, M., Alam, M. A., Benezra, A., DeStefano, J., Meier, M. F., Muegge, B. D. et al. Persistent Gut Microbiota Immaturity in Malnourished Bangladeshi Children. *Nature* **2014**, *510* (7505), 417–421. https://doi.org/10.1038/nature13421.

52. Kumar, M., Ji, B., Babaei, P., Das, P., Lappa, D., Ramakrishnan, G., Fox, T. E., Haque, R., Petri, W. A., Bäckhed, F. et al. Gut Microbiota Dysbiosis Is Associated with Malnutrition and Reduced Plasma Amino Acid Levels: Lessons from Genome-Scale Metabolic Modeling. *Metab. Eng.* **2018**, *49*, 128–142. https://doi.org/10.1016/j.ymben.2018.07.018.

53. Siegel, R. L., Miller, K. D., Jemal, A. Cancer Statistics, 2019. *CA. Cancer J. Clin.* **2019**, *69* (1), 7–34. https://doi.org/10.3322/caac.21551.

54. Freed-Pastor, W. A., Mizuno, H., Zhao, X., Langerød, A., Moon, S.-H., Rodriguez-Barrueco, R., Barsotti, A., Chicas, A., Li, W., Polotskaia, A. et al. Mutant P53 Disrupts Mammary Tissue Architecture via the Mevalonate Pathway. *Cell* **2012**, *148* (1–2), 244–258. https://doi.org/10.1016/j.cell.2011.12.017.

55. Nomura, D. K., Long, J. Z., Niessen, S., Hoover, H. S., Ng, S.-W., Cravatt, B. F. Monoacylglycerol Lipase Regulates a Fatty Acid Network That Promotes Cancer Pathogenesis. *Cell* **2010**, *140* (1), 49–61. https://doi.org/10.1016/j.cell.2009.11.027.

56. Joyce, J. A., Pollard, J. W. Microenvironmental Regulation of Metastasis. *Nat. Rev. Cancer* **2009**, *9* (4), 239–252. https://doi.org/10.1038/nrc2618.

57. Chang, S.-H., Liu, C. H., Conway, R., Han, D. K., Nithipatikom, K., Trifan, O. C., Lane, T. F., Hla, T. Role of Prostaglandin E2-Dependent Angiogenic Switch in Cyclooxygenase 2-Induced Breast Cancer Progression. *Proc. Natl. Acad. Sci. U. S. A.* **2004**, *101* (2), 591–596. https://doi.org/10.1073/pnas.2535911100.

58. Mahmud, I., Garrett, T. J., Liao, D. Abstract LB-266: DAXX Promotes de Novo Lipogenesis in Triple-Negative Breast Cancer. *Cancer Res.* **2017**, *77* (13 Supplement), LB-266-LB-266. https://doi.org/10.1158/1538-7445.AM2017-LB-266.

59. Mahmud, I., Liao, D. DAXX in Cancer: Phenomena, Processes, Mechanisms and Regulation. *Nucleic Acids Res.* **2019**, *47* (15), 7734–7752. https://doi.org/10.1093/nar/gkz634.

60. Mahmud, I., Garrett, T. J., Stowe, R., Roush, W. R., Liao, D. Abstract LB-248: Chemically Distinct Class I HDAC Inhibitors Synergize to Inhibit Global Lipid Metabolism in Cancer. *Cancer Res.* **2018**, *78* (13 Supplement), LB-248-LB-248. https://doi.org/10.1158/1538-7445.AM2018-LB-248.

61. Gouw, A. M., Margulis, K., Liu, N. S., Raman, S. J., Mancuso, A., Toal, G. G., Tong, L., Mosley, A., Hsieh, A. L., Sullivan, D. K. et al. The MYC Oncogene Cooperates with Sterol-Regulated Element-Binding Protein to Regulate Lipogenesis Essential for Neoplastic Growth. *Cell Metab.* **2019**, *30* (3), 556–572.e5. https://doi.org/10.1016/j.cmet.2019.07.012.

62. Clendening, J. W., Pandyra, A., Boutros, P. C., Ghamrasni, S. E., Khosravi, F., Trentin, G. A., Martirosyan, A., Hakem, A., Hakem, R., Jurisica, I. et al. Dysregulation of the Mevalonate Pathway Promotes Transformation. *Proc. Natl. Acad. Sci.* **2010**, *107* (34), 15051–15056. https://doi.org/10.1073/pnas.0910258107.

63. Bauer, D. E., Hatzivassiliou, G., Zhao, F., Andreadis, C., Thompson, C. B. ATP Citrate Lyase Is an Important Component of Cell Growth and Transformation. *Oncogene* **2005**, *24* (41), 6314–6322. https://doi.org/10.1038/sj.onc.1208773.

64. Hirsch, H. A., Iliopoulos, D., Joshi, A., Zhang, Y., Jaeger, S. A., Bulyk, M., Tsichlis, P. N., Shirley Liu, X., Struhl, K. A Transcriptional Signature and Common Gene Networks Link Cancer with Lipid Metabolism and Diverse Human Diseases. *Cancer Cell* **2010**, *17* (4), 348–361. https://doi.org/10.1016/j.ccr.2010.01.022.

65. Tuncer, S., Banerjee, S. Eicosanoid Pathway in Colorectal Cancer: Recent Updates. *World J. Gastroenterol.* **2015**, *21* (41), 11748–11766. https://doi.org/10.3748/wjg.v21.i41.11748.

66. Mills, G. B., Moolenaar, W. H. The Emerging Role of Lysophosphatidic Acid in Cancer. *Nat. Rev. Cancer* **2003**, *3* (8), 582–591. https://doi.org/10.1038/nrc1143.

67. Nelson, E. R., Wardell, S. E., Jasper, J. S., Park, S., Suchindran, S., Howe, M. K., Carver, N. J., Pillai, R. V., Sullivan, P. M., Sondhi, V. et al. 27-Hydroxycholesterol Links Hypercholesterolemia and Breast Cancer Pathophysiology. *Science* **2013**, *342* (6162), 1094–1098. https://doi.org/10.1126/science.1241908.

68. Yamada, T., Sato, K., Komachi, M., Malchinkhuu, E., Tobo, M., Kimura, T., Kuwabara, A., Yanagita, Y., Ikeya, T., Tanahashi, Y. et al. Lysophosphatidic Acid (LPA) in Malignant Ascites Stimulates Motility of Human Pancreatic Cancer Cells through LPA1. *J. Biol. Chem.* **2004**, *279* (8), 6595–6605. https://doi.org/10.1074/jbc.M308133200.

69. Morad, S. A. F., Cabot, M. C. Ceramide-Orchestrated Signalling in Cancer Cells. *Nat. Rev. Cancer* **2013**, *13* (1), 51–65. https://doi.org/10.1038/nrc3398.

70. Pyne, N. J., El Buri, A., Adams, D. R., Pyne, S. Sphingosine 1-Phosphate and Cancer. *Adv. Biol. Regul.* **2018**, *68*, 97–106. https://doi.org/10.1016/j.jbior.2017.09.006.

71. Park, J. B., Lee, C. S., Jang, J.-H., Ghim, J., Kim, Y.-J., You, S., Hwang, D., Suh, P.-G., Ryu, S. H. Phospholipase Signalling Networks in Cancer. *Nat. Rev. Cancer* **2012**, *12* (11), 782–792. https://doi.org/10.1038/nrc3379.

72. Yamaguchi, H., Wyckoff, J., Condeelis, J. Cell Migration in Tumors. *Curr. Opin. Cell Biol.* **2005**, *17* (5), 559–564. https://doi.org/10.1016/j.ceb.2005.08.002.

73. Friedl, P., Wolf, K. Tumour-Cell Invasion and Migration: Diversity and Escape Mechanisms. *Nat. Rev. Cancer* **2003**, *3* (5), 362–374. https://doi.org/10.1038/nrc1075.
74. Kloudova, A., Guengerich, F. P., Soucek, P. The Role of Oxysterols in Human Cancer. *Trends Endocrinol. Metab.* **2017**, *28* (7), 485–496. https://doi.org/10.1016/j.tem.2017.03.002.
75. Hannun, Y. A., Obeid, L. M. Principles of Bioactive Lipid Signalling: Lessons from Sphingolipids. *Nat. Rev. Mol. Cell Biol.* **2008**, *9* (2), 139–150. https://doi.org/10.1038/nrm2329.
76. Nishida, N., Yano, H., Nishida, T., Kamura, T., Kojiro, M. Angiogenesis in Cancer. *Vasc. Health Risk Manag.* **2006**, *2* (3), 213–219.
77. Nagahashi, M., Ramachandran, S., Kim, E. Y., Allegood, J. C., Rashid, O. M., Yamada, A., Zhao, R., Milstien, S., Zhou, H., Spiegel, S. et al. Sphingosine-1-Phosphate Produced by Sphingosine Kinase 1 Promotes Breast Cancer Progression by Stimulating Angiogenesis and Lymphangiogenesis. *Cancer Res.* **2012**, *72* (3), 726–735. https://doi.org/10.1158/0008-5472.CAN-11-2167.
78. Talmadge, J. E., Fidler, I. J. AACR Centennial Series: The Biology of Cancer Metastasis: Historical Perspective. *Cancer Res.* **2010**, *70* (14), 5649–5669. https://doi.org/10.1158/0008-5472.CAN-10-1040.
79. Massagué, J., Obenauf, A. C. Metastatic Colonization by Circulating Tumour Cells. *Nature* **2016**, *529* (7586), 298–306. https://doi.org/10.1038/nature17038.
80. Nieman, K. M., Kenny, H. A., Penicka, C. V., Ladanyi, A., Buell-Gutbrod, R., Zillhardt, M. R., Romero, I. L., Carey, M. S., Mills, G. B., Hotamisligil, G. S. et al. Adipocytes Promote Ovarian Cancer Metastasis and Provide Energy for Rapid Tumor Growth. *Nat. Med.* **2011**, *17* (11), 1498–1503. https://doi.org/10.1038/nm.2492.
81. Thomas, A., Patterson, N. H., Marcinkiewicz, M. M., Lazaris, A., Metrakos, P., Chaurand, P. Histology-Driven Data Mining of Lipid Signatures from Multiple Imaging Mass Spectrometry Analyses: Application to Human Colorectal Cancer Liver Metastasis Biopsies. *Anal. Chem.* **2013**, *85* (5), 2860–2866. https://doi.org/10.1021/ac3034294.
82. Komachi, M., Sato, K., Tobo, M., Mogi, C., Yamada, T., Ohta, H., Tomura, H., Kimura, T., Im, D.-S., Yanagida, K. et al. Orally Active Lysophosphatidic Acid Receptor Antagonist Attenuates Pancreatic Cancer Invasion and Metastasis *in Vivo. Cancer Sci.* **2012**, *103* (6), 1099–1104. https://doi.org/10.1111/j.1349-7006.2012.02246.x.
83. Pascual, G., Avgustinova, A., Mejetta, S., Martín, M., Castellanos, A., Attolini, C. S.-O., Berenguer, A., Prats, N., Toll, A., Hueto, J. A. et al. Targeting Metastasis-Initiating Cells through the Fatty Acid Receptor CD36. *Nature* **2017**, *541* (7635), 41–45. https://doi.org/10.1038/nature20791.
84. Yue, S., Li, J., Lee, S.-Y., Lee, H. J., Shao, T., Song, B., Cheng, L., Masterson, T. A., Liu, X., Ratliff, T. L. et al. Cholesteryl Ester Accumulation Induced by PTEN Loss and PI3K/AKT Activation Underlies Human Prostate Cancer Aggressiveness. *Cell Metab.* **2014**, *19*(3), 393–406. https://doi.org/10.1016/j.cmet.2014.01.019.
85. Cao, Y. Adipocyte and Lipid Metabolism in Cancer Drug Resistance. https://www.jci.org/articles/view/127201/pdf (accessed September 26, 2019). https://doi.org/10.1172/JCI127201.
86. Mahmud, I., Kabir, M., Haque, R., Garrett T. J. Decoding the Metabolome and Lipidome of Child Malnutrition by Mass Spectrometric Techniques: Present Status and Future Perspectives. *Anal. Chem.* **2019**, *91* (23), 14784–14791.
87. da Silva, R. R., Dorrestein, P. C., Quinn, R. A. Illuminating the Dark Matter in Metabolomics. *PNAS.* **2015**, *112* (41), 12549–12550.

3 Tracking and Imaging
SERS Characterization of Biomembranes

Wen Ren and Joseph Irudayaraj

CONTENTS

3.1 INTRODUCTION OF SERS FOR BIOMEMBRANE INVESTIGATION

Biomembrane is a fundamental component of most living systems. It is the structure separating the exterior regions of the intracellular components from the extracellular environment, yet selectively connecting the cell to the environment with materials and chemical signals. To understand the structure and function of biomembranes, identification of components, mapping their distribution and monitoring the related biological process is helpful and possible by SERS. Raman spectra are generated in response to molecular vibration due to polarizability change of molecules which could provide a wealth of information of the structures, enabling label-free recognition of targets. Compared to infrared spectroscopy, the Raman signal is not affected by the presence of water, facilitating characterization of biomembranes in solution phase. However, due to the limited Raman scattering cross section, the intensity of Raman spectrum is usually weak, thus influencing its signal. Surface-enhanced Raman spectroscopy (SERS) could greatly enhance the intensity of the spectrum with an average enhancement factor of ~10^6,[1] and perhaps as high as 10^{14}–10^{15},[2] as claimed in some reports, facilitating detection at lower concentrations and from finer structure changes.

To achieve a strong spectral intensity, SERS substrates based on SERS active materials such as gold, silver, and copper, etc., are necessary.[3–6] Morphologies such as sharp edges or tips of nanostructures, or gaps between nanoparticles that exhibit strong surface plasmonic resonance (SPR), or localized surface plasmonic resonance (LSPR), could provide extremely strong SERS activity due to the electromagnetic effect. These nanoscale structures are termed as "hot spots."[7–9] The enhanced signal reduces the detection time, making it possible to monitor changes in molecular structures and membrane dynamics in real time.

The usage of SERS substrates enables unique strategies for the investigation of bioactive membranes. With SERS, the enhancement of SERS substrates is localized in a range near the surface of substrates, usually within several nanometers. Therefore, in spectroscopic investigation, only the spectra from the biomembrane near the substrates could be recognized whereas the signal from the material away from the substrates is minimal. The low amount and similar chemical structure of the molecules in biomembrane makes it hard to differentiate the molecules. SERS probes could label the particular target molecules along with the corresponding SERS spectrum. The modification of Raman reporters could provide stronger signals to represent the existence of target molecules, and their unique fingerprint peaks facilitate multiplex identification of components in biomembrane.

SERS imaging has been performed with mammalian cells, bacteria cells, even zebrafish embryos and mice.[10–16] Briefly, there are two types of SERS imaging strategies: (i) the SERS spectra of all the components enhanced by substrates are nonselectively collected and the image is constructed based on the spectrum from each pixel; (ii) target molecules are selectively enhanced or labeled, and only the spectra from the target molecules or the labeled probe are recorded in the image. Such approaches require a substrate with large area and more importantly uniform SERS activity, commonly with the fabrication of novel surfaces or nanoparticle arrays based on SERS active materials as the substrates. The strategy could offer sufficient information on the distribution of these components in membranes; however, the uniformity of SERS activity of the substrates may influence the determination of the amount of the component. The second method could be performed with different probes such as SERS active nanoparticles modified with recognition ligands and sometimes Raman reporters, or Raman reporters modified with recognition ligands. Since a feature signal comes from labeled target molecules or Raman reporters on probes, the resulting image could clearly display the distribution of the target. Furthermore, it should be noted that the SERS image is based on the spectra from the components of the biomembrane and has a low-signal to noise ratio due to the background noise or competing signals. Based on the signal from Raman reporters on the SERS probes, it is easier to differentiate similar elements in the image; however, since the signal intensity is determined by the reporters enhanced by nanoparticles, non-specific interaction and labeling efficiency of the probe-labeling process will influence the target images.

3.2 SERS FINGERPRINT-BASED INVESTIGATION OF BIOMEMBRANE

3.2.1 LABEL-FREE SERS OF BIOMEMBRANE

The SERS spectra of molecules could be used to directly identify the components on biomembrane. To probe the fundamental structure of biomembrane, investigation of lipid bilayers or multilayers would be a good start. Utilizing Ag electrodes or Ag coated substrates, the SERS spectra from lipid layers were obtained.[17,18] With a L-α-dipalmitoylphosphatidylcholine (DPPC) bilayer, the SERS spectra, especially the ratio of 1105 cm^{-1} (*gauche* vibrational model of DPPC) to 1142 cm^{-1} (trans vibrational model of DPPC), was used to indicate the order-to-disorder phase transition with respect to voltage or temperatures.[17] Further, the SERS spectra from the lipid layer could differentiate the lipid bilayers from monolayers.[18] These initial efforts inform on the interrogation of lipid bilayers with SERS. In such experiments, the peak of the headgroup symmetric stretch and double bond stretch in the SERS spectra of lipids could reveal the position and orientation of lipids in the bilayers.[19] Beyond the identification of lipid molecules and characterization of lipid bilayer structure, SERS spectra could further reveal the dynamics of lipid molecules on the lipid bilayer. Halas and colleagues constructed 1,2-dimyristoyl-sn-glycero-3-phosphocholine (DMPC) lipid layer on gold nanoshells and donor-deuterated DMPC vesicles on the DMPC layer.[20] The gold nanoshells as SERS substrates provided localized enhancement of the lipid layer close to the shell surface. Since the vibrational fingerprint of C-D stretch and C-H stretch mode are different, the transfer of deuterated DMPC from vesicles to the lipid layer near gold shell would induce a change in SERS spectra. Thus, with the corresponding SERS peaks, the transfer of deuterated DMPC could be monitored in real-time. Similarly with deuterated lipid molecules and gold nanorods as SERS substrates, the SERS spectra of lipid molecules could also differentiate between a bilayer and nonbilayer with respect to lipid vesicle concentration in solution.[21] It should also be noted that the difference in lipid bilayer composition does not always induce a change in the SERS spectra. For instance, Chazalet and colleagues found that the difference in SERS spectra from palmitoyl-oleoyl-phosphatidylcholine (POPC)-cholesterol bilayer and pure POPC bilayer is not significant.[18] Such indistinctive SERS spectra signatures, not only between the phospholipid molecules but also between other components of the biomembrane, would be one of the primary limitation in the SERS evaluation of biomembrane. As the primary components in biomembrane, with SERS it is easy to identify the lipid composition of biomembrane. However, to track the movement of lipids on membranes, highly active SERS substrates as well as suitably designed measurement strategy are necessary.

The research discussed above exhibited the capability of SERS to monitor small chemicals on biomembranes based on their spectral signatures, which could be extended to proteins on biomembrane. Proteins play an important role in the function of biomembrane such as material transport, signaling, and immune recognition. Such protein-related study is very valuable for the understanding of biomembranes. Although SERS signatures from proteins are not as distinct as that from small molecules because of the molecular structure, some efforts have been reported in this space. For instance, with biotinylated lipid bilayers and streptavidin, Grebel and colleagues characterized the binding of protein on lipid bilayer in a microfluidic system.[22] The binding of streptavidin was confirmed based on the presented SERS peak at 1410 cm^{-1} as the signature. Based on the change in SERS intensity, the influence from washing and the presence of large suspended cells (*Salmonella enterica* as model cells) and the interaction between streptavidin and biotinylated lipid bilayer were investigated. Beyond the binding of the protein to biomembrane, Maiti and colleagues utilized SERS to evaluate Amyloid-β40 (Aβ40), an oligomer related to Alzheimer's disease, to investigate its conformation on lipid bilayer.[23] Assisted with ^{13}C labeling, the peak shifts in Aβ40 SERS spectra were used to inform on the secondary structure of Aβ40 in the lipid bilayer. In this application, an *in-situ* route to detect the conformation of proteins on biomembranes by SERS was demonstrated. The SERS based strategy could be extended to characterizing proteins on cell membranes. Manno and colleagues fabricated AuNPs as SERS substrates to investigate proteins on cell membranes.[24] The target protein PrP^C interacting with copper ions could increase the SERS intensity in the 1500–1800 cm^{-1} range. More specifically, the ratio between 1577 and 1603 cm^{-1} was used as a signature for copper bound PrP^C. The signature, after the interaction between copper ions and PrP^C, exhibited different values upon PrP^C expression level in four cell models. The change in the signature upon the PrP^C expression suggests a potential route to *in-situ* monitoring of protein expression on cell membrane. Lee et al. used SERS enhancement to quantify cell surface receptors and validated with plasmon imaging.[25]

Beyond proteins, 10–200 nm lipid raft domains in plasma membrane of cells with higher density of sphingomyelin and cholesterol, has been demonstrated in relation with functions of protein/lipid sorting and signaling. Mas-Oliva and colleagues utilized SERS of fractions from plasma membrane of hepatocyte to investigate lipid rafts.[26] Principal component analysis (PCA) was used to analyze the SERS spectra as indicators of lipid rafts. This research showed a correlation between lipid rafts and catalytic activity of Ca^{2+}-ATPase.

In addition to lipid molecules, other materials could be recognized based on their SERS signatures, thus enabling research on the role of these materials on lipid bilayers. For instance, with 1,2-dioleoyl-sn-glycero-3-phosphocholine (DOPC) bilayer encapsulated silver nanoparticles (AgNPs), Rusciano and colleagues investigated the penetration of ultra-fine nano-sized organic carbon particles (NOCs, 1–5 nm) in lipid bilayers.[27] After around 10 minutes of addition of NOCs to the solution of DOPC bilayer wrapped AgNPs, the obtained NOCs SERS spectra indicated that the NOCs permeated the lipid bilayer to the surface of AgNPs. Our results implied that beyond the common endocytotic transmembrane diffusion, direct penetration could be another possible route for the NOCs passing through the lipid bilayer-based biomembrane in a short time. Benefiting from the advancements in nanotechnology, it is also possible to manufacture SERS substrates allowing to track the molecular movement along the lipid layers. Lohmüller and colleagues prepared plasmonic nanoantennas array with gold nanotriangles as SERS substrate.[28] The gaps between the tips of the nanotriangles could provide strong SERS enhancement due to LSP resonance. The lipid bilayers were constructed between gold nanotriangles, when target molecules on lipid layer passed along the layer through these gaps to yield a corresponding SERS spectra.

Based on their feature SERS spectra, the investigations on chemical drugs interaction with lipid bilayers or biomembranes were also reported. It is noted that permeability of drugs across biomembrane is related to their efficacy and cytotoxicity.[29,30] SERS could provide a better understanding of drug permeation across biomembranes. Because of the cytotoxicity, research on the dynamics of antitumor drugs on membranes is expected to improve treatment efficacy while reducing intracellular accumulation with optimized influx and efflux. Chazalet and colleagues initiated experiments on the behaviors of anthracyclines, chemicals used in chemotherapy, from pirarubicin and to more anthracyclines.[31,32] Based on the SERS spectra of anthracyclines and lipid layers, the orientation of anthracyclines in lipid bilayer could be uncovered. Furthermore, based on changes in SERS spectra the difference in behavior of the three anthracyclines was confirmed: pirarubicin could cross the lipid bilayers based on dipalmitoylphosphatidic acid (DPPA) or POPC-DPPA on the surface, while Adriamycin, another anthracycline, would be adsorbed in the second monolayer and thus could not cross the bilayer in the presence of DPPA. Per the composition of lipid bilayers, the behavior of different chemical drugs would be different. Further details will be uncovered in the following work with DPPA, an anionic phospholipid, in a first layer of lipid bilayers could limit the pirarubicin permeation through lipid bilayers.[33] Millot and colleagues moved the research of drug behavior from lipid bilayer model to the plasma membrane in living cells. They first investigated the adsorption of mitoxantrone (MTX), an antitumor drug, on the plasma membrane of MTX-sensitive and MTX-resistant cells.[34] The SERS intensity of MTX showed the adsorption of MTX on both MTX-sensitive cells and MTX-resistant cells. Further, they investigated the adsorption and permeability of MTX on resistant cells under serial conditions.[35]

To further enrich the information from SERS, other analytic techniques were introduced and combined with SERS. Infrared spectrum, induced with the vibration with dipole change, could provide complementary information of SERS spectra and was used for SERS evaluation. For instance, reflection absorption infrared spectroscopy and surface enhanced infrared spectroscopy were combined with SERS to study lipid bilayers.[36-38] Because of the non-uniform distribution of components on biomembrane, the integrated spectral and topographical information obtained from atomic force microscopy (AFM) was used to evaluate the chemical distribution as well as the topography of the mimetic lipid bilayer.[39] Lesniewska and colleagues used this combined methodology to study murine norovirus binding to cell membranes, and found the relation between lipid rafts on cell membrane and the infection process.[40] With SERS active material-based electrodes or nanostructures, it is possible to combine SERS with electrochemical strategies. The electrochemically enhanced SERS methods have been utilized in various membrane evaluation studies: the interaction between protein and lipid membrane,[41,42] activity of different enzymes on lipid bilayers,[43-47] the micro-structure of lipid bilayers,[48] and electron-transfer process in lipid bilayers.[49] The combination of SERS with other analytic techniques are promising routes for further investigation of biomembrane.

3.2.2 SERS PROBES FOR BIOMEMBRANE ANALYSIS

It is understood that direct observation of targets on biomembrane is the most desirable. However, not all targets could provide distinct SERS signatures, while the complexity of biomembrane makes it hard to differentiate molecules with similar chemical structures. Alternatively, with appropriate labeling, SERS probes are able to differentiate biomolecules in similar structures,[50,51] which could also be used for the study of biomembrane. For instance, Olivo and colleagues used 4-(dihydroxyborophenyl) acetylene (DBA), a Raman reporter, as probes that could specifically interact with sialic acid to detect sialic acid on cell membrane.[52] The strong signal from DBA could determine the existence of sialic acid on cell membrane. Similarly, Liang and colleagues prepared SERS probes of AgNPs modified with 4-mercaptophenylboronic acid (MPBA), which could yield strong SERS spectra to interact with target sialoglycan expression on cell membrane.[53] The SERS spectra of MPBA changes upon the binding with target sialoglycan, which could be used as an indicator of sialoglycan. The interaction of MPBA with sialic acid was also utilized for SERS imaging of sialic acid distribution on cancer cells.[54] More commonly, the identification of target molecules on biomembrane was carried out with recognition ligands modified on the SERS probes. By integrating antibody and reporter molecules on AuNPs, Coronado and colleagues detected metabotropic glutamate receptor 1a, a neuronal cell membrane receptors, expressed in neuronal cells.[55] Bamrungsap and colleagues conjugated aptamer and 4-aminothiophenol (4-ATP), a Raman reporter, to Au nanorods to fabricate SERS probes to detect human protein tyrosine kinase-7 (PTK-7) on Hela cell membrane.[56] Based on SERS signal of 4-ATP, the expression of PTK-7 on cell membrane could be determined. Beyond specific recognition, some labeling strategies could achieve "semi-recognition" of particular groups of biomolecules on biomembrane. For instance, utilizing the interaction between Ag^+ ions and phosphatidylserine, Chen and colleagues investigated apoptosis based on SERS from cell membrane.[57] For normal cells, the phosphatidylserine was found to be mostly distributed in the inter leaflet of the cell membrane, thus Ag^+ ions will not attach to the cells; for apoptotic cells, the phosphatidylserines were externalized to interact with Ag^+ ions due to the electrostatic effect, which was reduced to AgNPs in the presence of reducing agents to enhance the SERS signal as an indicator of apoptosis. Unlike the strategy with nanoparticles generated in cell,[58] this method could selectively enhance the SERS signal from cell membrane for the investigation of the apoptotic process. Marunaka and colleagues conjugated AuNPs to proteins on cell membrane with sulfo-NHS-ester-biotin linker and streptavidin-biotin interaction.[59] 4-mercaptobenzoic acid as Raman reporter was modified on AuNPs which could respond to pH change. With this technique, the pH near the proteins on cell membrane was characterized and SERS mapping was performed to illustrate the pH distribution around cell membranes. As discussed above, SERS investigation based on probe-labeling strategy could confirm the presence of specific target molecules on biomembranes, while the multi-functionalized probes could enable the detection of targets and provide other information such as pH, at particular regions of biomembranes. However, some limitations with this approach are: (i) non-specific interaction between modified probes and biomembranes and therefore a false signal is possible; (ii) labeling efficiency of probes to the target on biomembranes, as well as some components may be located on the inner leaflets; (iii) the influence of probe-labeling on the nature of biomembrane.

3.3   SERS IMAGING ON BIOMEMBRANE

Based on the SERS spectra collected from target components, label-free SERS images could be constructed. One of the common substrates is the array prepared with SERS active nanoparticles for label-free SERS imaging. For instance, Wang and colleagues prepared AgNPs arrays based on a liquid-liquid interface assembly strategy and then casted lipid membrane based on 1,2-dimyristoyl-sn-lycero-3-phosphoglycerol, sodium salt (DMPG) and DMPC.[60] The SERS of DMPC and DMPG on the substrates was collected and the feature peak at 1482 cm^{-1} was used to identify DMPG. To further reduce the influence from the change in SERS activity of the substrate, the ratio of 1482 to 1650 cm^{-1}, a peak in both DMPC and DMPG, was used to present the ratio of the DMPG amount to DMPC in the lipid membrane. The resulting image exhibited non-uniform distribution of negatively charged DMPG on the membrane, which paved a possible route to the investigation of microstructure such as lipid raft on biomembrane. Beyond the distribution of specific lipids, the SERS spectra could reveal the status of lipid bilayer. Mori and colleagues casted 1,2-distearoyl-sn-glycero-3-phosphocholine (DSPC) based lipid bilayer on silver covered glass and constructed a SERS mapping strategy based on the spectra of DSPC.[61] The ratio of CH$_3$ asymmetric stretching at 2962 cm^{-1} to CH$_2$ symmetric stretching at 2849 cm^{-1} indicated the randomly stacked and densely packed lipid bilayer, the corresponding mapping illustrated the area of lipid bilayer in different status. Instead of lipid bilayer as the mimetic biomembrane, Tian and colleagues placed Chinese hamster lung fibroblast cells on AuNP array formed on ITO glass and collected the SERS image of the cell membrane.[62] As shown in Figure 3.1, the Raman signal enhanced spectra with multiple peaks varied at different spots of the cell membrane. Four signature peaks were chosen to indicate the amino acids of phenylalanine, cysteine, proline and methionine, and SERS images based on these peaks were constructed to present the distribution of these amino acids on cell membrane. Similar concept was applied with red blood cell by Zito and colleagues with nano-textured AgNP array.[63,64] SERS images obtained based on the typical band of protein group Amide I or the integrated intensity from 1100 to 1700 cm^{-1}. Both of the images showed strong correlation to the optical morphology of red blood cells. Beyond the nanoparticle arrays, the advance of nanotechnology could provide new substrates for SERS imaging. Gracias and colleagues used quartz wafer to prepare mechanical trap, and on the inner surface of the wafer Au nanostars were modified for enhanced SERS imaging of MDA-MB-231 breast cancer cells.[65] According to the discussion above, it can be seen that the SERS image with the spectra directly from membranes could provide enriched information in multiple-channel imaging. Meanwhile, it should be noted that the obtained images could possibly be influenced by SERS from non-target molecules on the membrane with similar SERS peak.

SERS imaging with probes is an alternative method to reduce the influence from indistinctive SERS spectra to achieve a clearer image of the target component on biomembrane. For instance, Pezacki and colleagues synthesized a chemical as a Raman reporter for imaging of platelet-derived growth factor receptor on Hela cells.[66] AgNPs modified with the chemicals were labeled on cells for SERS imaging.

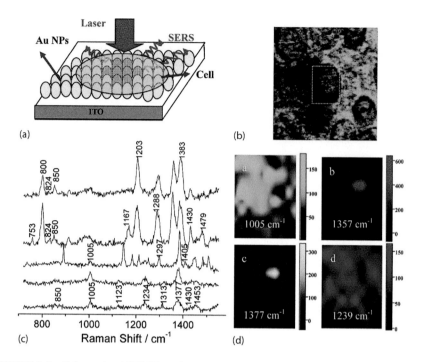

(a)   (b)   (c)   (d)

FIGURE 3.1 Schematic of SERS imaging of cell membrane on AuNP array (a), video image of cell (b), SERS spectra at different positions of cell (c) and corresponding SERS images based on peaks at 1005 cm^{-1} (Phe, a), 1357 cm^{-1} (Cys, b), 1377 cm^{-1} (Pro, c) and 1239 cm^{-1} (Met, d) (d). (Reprinted with permission from Mori, M. et al., *Biomedical Vibrational Spectroscopy 2018: Advances in Research and Industry*, SPIE BiOS, San Francisco, CA, 2018. Copyright 2008 American Chemical Society.)

The research also demonstrated the cell membrane damage from nanoparticle labeling due to the photothermal effect. Popp and colleagues fabricated hydroxy-apatite, which could be linked to the carboxyl group on cell membrane to yield strong Raman signal, for imaging the carboxyl group distribution on carcinoma cell membrane.[67] SERS imaging with a confocal Raman system was carried out at different focal planes of cells. Chen and colleagues utilized the interaction of chemicals as probes to particular molecules on biomembrane to label the cell membrane for SERS images.[68] Due to the specific interaction, the SERS of the probe chemicals were used to label proteins, glycan, and choline-containing phospholipids, and the obtained SERS images could indicate the distribution of these components on cell membrane. Beyond the chemical interaction, cationic gold nanoparticles were used to label glycocalyx on cell membrane.[69] Compared to these "semi-specific" labeling strategies, more specific labeling with probes modified with recognition ligands such as antibody or aptamer could be also used for SERS imaging. Besides the aptamer based SERS imaging discussed above,[56] antibody modified nanoparticles have been used to label keratan sulfate on endothelial cell membrane to enable SERS imaging.[70] As shown in Figure 3.2, we modified AuNPs with antibody to label CD44

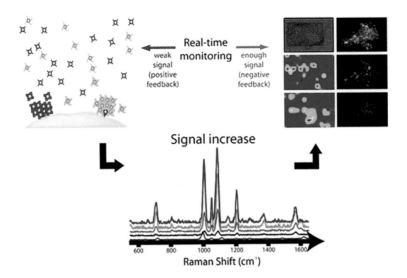

Signal increase

Raman Shift (cm⁻¹)

FIGURE 3.2 Antibody-based probe recognition and improved SERS for cancer cell surface marker detection and SERS imaging. (Reprinted with permission from Hodges, M. D. et al., *ACS Nano*, 5, 9535–9541, 2011. Copyright 2011 American Chemical Society.)

and CD24, cell surface markers on different breast cancer cell lines.[25] Enhanced with DNA modified AuNPs, SERS images demonstrate the expression of surface markers and their distribution. Instead of the Raman reporter, the SERS spectra from biomolecules near the labeled nanoparticles reveal the existence of lipid and protein around keratan sulfate on cell membrane. Compared with label-free SERS imaging, images with probe labeling provides more specific and distinct images of the target components on biomembrane. However, limitations of SERS imaging with probes should take into account the photothermal damage due to the heat generated from the nanoparticles upon excitation, the labeling efficiency and non-specific binding as well as possible probe trapping due to the physiological activity of living cells.

3.4 TIP-ENHANCED RAMAN SCATTERING ON BIOMEMBRANE

Benefiting from the development of near-field optical technologies such as scanning near-field optical microscopy (SNOM), tip-enhanced Raman scattering (TERS) is accomplished, by integrating SERS and scanning probe microscopy (SPM), to provide nanoscale resolution imaging of proteins and lipids. In TERS, a metalized tip with a sharp top could offer a highly localized SERS enhancement near the top to collect spectra from the chemicals in the confined range; the measurement of SPM, for example AFM, would record high-resolution spatial information simultaneously. The combination of SERS spectra and the corresponding position could result in a SERS-based image. TERS is an excellent technique for the investigation of membranes due to the high resolution and corresponding SERS information from the components of the membrane at nanometer precision. In one of the initial explorations, Popp and colleagues used TERS to characterize lipid layers

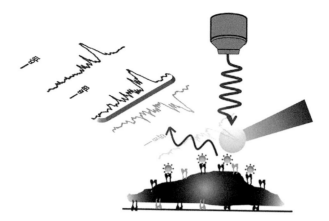

FIGURE 3.3 Schematic of selective TERS with AuNP probe labeling for the recognition of cell surface marker. (Reprinted with permission from Xiao, L. et al., *Anal. Chem.*, **88**, 6547–6553, 2016. Copyright 2008 American Chemical Society.)

and cell membranes.[71] They also evaluated protein-containing lipid membranes at very high resolutions not possible with conventional SERS.[72] With lipid monolayers, Zenobi and colleagues performed AFM-SERS image of lipid layer to elucidate the lipid domains in the layers.[73] The ratio between peak intensity at 2100 and 2900 cm^{-1} in the image with 128 × 128 pixels is illustrated in the lipid domain with a higher ratio of d62-DPPC in the lipid layer. TERS was used to investigate the distribution of different components on cell membranes. Based on the spectra from inherent molecules on the membranes, Deckert and colleagues demonstrated a non-uniform distribution of lipids and proteins on cell membranes.[74] Probe-labeling strategy could be also used for TERS imaging. As shown in Figure 3.3, Schultz and colleagues incubated cells with ligand-modified AuNPs, which could recognize various integrins on cell membranes, and the collected TERS image to present the corresponding target distribution on cell membranes,[75,76] demonstrating a selective TERS imaging method for the investigation of specific targets on biomembranes. Based on the past work, we note that TERS is a promising technique for biomembrane investigation both for single point and label-free as well as high-resolution imaging.

3.5 CONCLUSION

In this brief review, we discussed the application of SERS to evaluate biomembranes, including the capabilities of SERS for the identification of various biomolecules, tracking its movement and conformational changes of the components based on composition, function and state of the membrane. Information from such investigations will help in the understanding of the function of a biomembrane and its response to chemicals and environment changes. We propose that SERS could be a promising technology for the investigation of biomembranes due to its unique capabilities, multiplexing ability, highly resolved spectrum, and enhanced sensitivity. Furthermore, SERS could be combined with other analytic techniques such as electrochemical methods, AFM and other spectral methods to provide complementary quantitative or structural information. Information from such integrated systems could provide complementary data on the state of the analyte or substrate. Meanwhile, SERS based imaging as well as TERS could provide spectra with a high spatial resolution, which could be used for mapping the distribution and movement of target components. Accordingly, the potential of SERS for multicomponent imaging and elucidation of chemical structures in membranes makes it a compelling analytical tool in this field.

REFERENCES

1. Fan, M., Andrade, G. F., Brolo, A. G. *Anal. Chim. Acta* **2011**, *693*, 7–25.
2. Nie, S., Emory, S. R. *Science* **1997**, *275*, 1102–1106.
3. Vo-Dinh, T. *TrAC, Trends Anal. Chem.* **1998**, *17*, 557–582.
4. Hildebrandt, P., Stockburger, M. *J. Phys. Chem.* **1984**, *88*, 5935–5944.
5. Wang, Y., Asefa, T. *Langmuir* **2010**, *26*, 7469–7474.
6. Lee, K., Irudayaraj, J. *Small* **2013**, *9*, 1106–1115.
7. Willets, K. A., Van Duyne, R. P. *Annu. Rev. Phys. Chem.* **2007**, *58*, 267–297.
8. Tian, Z. Q., Ren, B., Wu, D. Y. *J. Phys. Chem. B* **2002**, *106*, 9463–9483.
9. Ouyang, L., Hu, Y., Zhu, L., Cheng, G. J., Irudayaraj, J. *Biosens. Bioelectron.* **2017**, *92*, 755–762.
10. Song, J. B., Zhou, J. J., Duan, H. W. *J. Am. Chem. Soc.* **2012**, *134*, 13458–13469.
11. Yu, C., Gestl, E., Eckert, K., Allara, D., Irudayaraj, J. *Cancer Detect. Prev.* **2006**, *30*, 515–522.
12. Lee, S., Kim, S., Choo, J., Shin, S. Y., Lee, Y. H., Choi, H. Y., Ha, S. H., Kang, K. H., Oh, C. H. *Anal. Chem.* **2007**, *79*, 916–922.
13. Ravindranath, S. P., Henne, K. L., Thompson, D. K., Irudayaraj, J. *ACS Nano* **2011**, *5*, 4729–4736.
14. Ravindranath, S. P., Henne, K. L., Thompson, D. K., Irudayaraj, J. *PLoS One* **2011**, *6*, e16634.
15. Wang, Y., Seebald, J. L., Szeto, D. P., Irudayaraj, J. *ACS Nano* **2010**, *4*, 4039–4053.
16. Zavaleta, C. L., Smith, B. R., Walton, I., Doering, W., Davis, G., Shojaei, B., Natan, M. J., Gambhir, S. S. *Proc. Natl. Acad. Sci. U. S. A.* **2009**, *106*, 13511–13516.
17. Guo, F. C., Chou, Y. C., Huang, W. N., Wu, W. G. *J. Raman Spectrosc.* **1992**, *23*, 425–430.
18. Chazalet, M. S., Masson, M., Bousquet, C., Bolbach, G., Ridente, Y., Bolard, J. *Thin Solid Films* **1994**, *244*, 852–856.
19. Matthews, J. R., Shirazinejad, C. R., Isakson, G. A., Demers, S. M., Hafner, J. H. *Nano Lett.* **2017**, *17*, 2172–2177.
20. Kundu, J., Levin, C. S., Halas, N. J. *Nanoscale* **2009**, *1*, 114–117.
21. Matthews, J. R., Payne, C. M., Hafner, J. H. *Langmuir* **2015**, *31*, 9893–9900.
22. Banerjee, A., Perez–Castillejos, R., Hahn, D., Smirnov, A. I., Grebel, H. *Chem. Phys. Lett.* **2010**, *489*, 121–126.
23. Bhowmik, D., Mote, K. R., MacLaughlin, C. M., Biswas, N., Chandra, B., Basu, J. K., Walker, G. C., Madhu, P. K., Maiti, S. *ACS Nano* **2015**, *9*, 9070–9077.
24. Manno, D., Filippo, E., Fiore, R., Serra, A., Urso, E., Rizzello, A., Maffia, M. *Nanotechnology* **2010**, *21*, 165502.
25. Lee, K., Drachev, V. P., Irudayaraj, J. *ACS Nano* **2011**, *5*, 2109–2117.
26. Delgado-Coello, B., Montalvan-Sorrosa, D., Cruz-Rangel, A., Sosa-Garrocho, M., Hernández-Téllez, B., Macías-Silva, M., Castillo, R., Mas-Oliva, J. *J Raman Spectrosc.* **2017**, *48*, 659–667.

27. Rusciano, G., De Luca, A., Pesce, G., Sasso, A. *Carbon* **2009**, *47*, 2950–2957.
28. Kühler, P., Weber, M., Lohmüller, T. *ACS Appl. Mater. Interfaces* **2014**, *6*, 8947–8952.
29. Veldman, R., Zerp, S., Van Blitterswijk, W., Verheij, M. *Br. J. Cancer* **2004**, *90*, 917.
30. Hendrich, A., Michalak, K. *Curr. Drug Targets* **2003**, *4*, 23–30.
31. Heywang, C., Saint–Pierre Chazalet, M., Masson, M., Garnier–Suillerot, A., Bolard, J. *Langmuir* **1996**, *12*, 6459–6467.
32. Heywang, C., Saint-Pierre-Chazalet, M., Masson, M., Bolard, J. *Langmuir* **1997**, *13*, 5634–5643.
33. Heywang, C., Chazalet, M. S.-P., Masson, M., Bolard, J. *Biophys. J.* **1998**, *75*, 2368–2381.
34. Breuzard, G., Angiboust, J.-F., Jeannesson, P., Manfait, M., Millot, J.-M. *Biochem. Biophys. Res. Commun.* **2004**, *320*, 615–621.
35. Breuzard, G., Piot, O., Angiboust, J.-F., Manfait, M., Candeil, L., Del Rio, M., Millot, J.-M. *Biochem. Biophys. Res. Commun.* **2005**, *329*, 64–70.
36. Meuse, C. W., Niaura, G., Lewis, M. L., Plant, A. L. *Langmuir* **1998**, *14*, 1604–1611.
37. Leverette, C. L., Dluhy, R. A. *Colloids Surf. A: Physicochem. Eng. Asp.* **2004**, *243*, 157–167.
38. Levin, C. S., Kundu, J., Janesko, B. G., Scuseria, G. E., Raphael, R. M., Halas, N. J. *J. Phys. Chem. B* **2008**, *112*, 14168–14175.
39. Sweetenham, C. S., Larraona-Puy, M., Notingher, I. *Appl. Spectrosc.* **2011**, *65*, 1387–1392.
40. Aybeke, E. N., Belliot, G., Lemaire-Ewing, S., Estienney, M., Lacroute, Y., Pothier, P., Bourillot, E., Lesniewska, E. *Small* **2017**, *13*, 1600918.
41. Karaballi, R. A., Merchant, S., Power, S. R., Brosseau, C. L. *PCCP* **2018**, *20*, 4513–4526.
42. Millo, D., Bonifacio, A., Moncelli, M. R., Sergo, V., Gooijer, C., van der Zwan, G. *Colloids Surf. B: Biointerfaces* **2010**, *81*, 212–216.
43. Friedrich, M. G., Gieβ, F., Naumann, R., Knoll, W., Ataka, K., Heberle, J., Hrabakova, J., Murgida, D. H., Hildebrandt, P. *Chem. Commun.* **2004**, 2376–2377.
44. Murgida, D. H., Hildebrandt, P. *PCCP* **2005**, *7*, 3773–3784.
45. Hrabakova, J., Ataka, K., Heberle, J., Hildebrandt, P., Murgida, D. H. *PCCP* **2006**, *8*, 759–766.
46. Todorovic, S., Verissimo, A., Wisitruangsakul, N., Zebger, I., Hildebrandt, P., Pereira, M. M., Teixeira, M., Murgida, D. H. *J. Phys. Chem. B* **2008**, *112*, 16952–16959.
47. Friedrich, M. G., Robertson, J. W., Walz, D., Knoll, W., Naumann, R. L. *Biophys. J.* **2008**, *94*, 3698–3705.
48. Vezvaie, M., Brosseau, C. L., Lipkowski, J. *Electrochim. Acta* **2013**, *110*, 120–132.
49. Ma, W., Ying, Y.-L., Qin, L.-X., Gu, Z., Zhou, H., Li, D.-W., Sutherland, T. C., Chen, H.-Y., Long, Y.-T. *Nat. Protoc.* **2013**, *8*, 439.
50. Sun, L., Yu, C., Irudayaraj, J. *Anal. Chem.* **2007**, *79*, 3981–3988.
51. Sun, L., Yu, C., Irudayaraj, J. *Anal. Chem.* **2008**, *80*, 3342–3349.
52. Gong, T. X., Cui, Y., Goh, D., Voon, K. K., Shum, P. P., Humbert, G., Auguste, J. L., Dinh, X. Q., Yong, K. T., Olivo, M. *Biosens. Bioelectron.* **2015**, *64*, 227–233.
53. Liang, L. J., Qu, H. X., Zhang, B. Y., Zhang, J., Deng, R., Shen, Y. T., Xu, S. P., Liang, C. Y., Xu, W. Q. *Biosens. Bioelectron.* **2017**, *94*, 148–154.
54. Liang, L., Shen, Y., Zhang, J., Xu, S., Xu, W., Liang, C., Han, B. *Anal. Chim. Acta* **2018**, *1033*, 148–155.
55. Fraire, J. C., Masseroni, M. L., Jausoro, I., Perassi, E. M., Diaz Añel, A. M., Coronado, E. A. *ACS Nano* **2014**, *8*, 8942–8958.
56. Bamrungsap, S., Treetong, A., Apiwat, C., Wuttikhun, T., Dharakul, T. *Microchim. Acta* **2016**, *183*, 249–256.
57. Zhou, H. B., Wang, Q. Q., Yuan, D. T. et al. *Analyst* **2016**, *141*, 4293–4298.

58. Shamsaie, A., Jonczyk, M., Sturgis, J. D., Robinson, J. P., Irudayaraj, J. *J. Biomed. Opt.* **2007**, *12*, 020502.

59. Puppulin, L., Hosogi, S., Sun, H. X., Matsuo, K., Inui, T., Kumamoto, Y., Suzaki, T., Tanaka, H., Marunaka, Y. *Nat. Commun.* **2018**, *9*.

60. Ren, W., Liu, J., Guo, S., Wang, E. *Sci. China Chem.* **2011**, *54*, 1334.

61. Mori, M., Abe, S., Kondo, T., Saito, Y. *Biomedical Vibrational Spectroscopy 2018: Advances in Research and Industry*, SPIE BiOS, San Francisco, CA, 2018.

62. Li, M.-D., Cui, Y., Gao, M.-X., Luo, J., Ren, B., Tian, Z.-Q. *Anal. Chem.* **2008**, *80*, 5118–5125.

63. Zito, G., Malafronte, A., Dochshanov, A., Rusciano, G., Auriemma, F., Pesce, G., De Rosa, C., Sasso, A. *Optical Sensors 2013*, International Society for Optics and Photonics, 2013, p. 87740B. Bellingham, Washington.

64. Zito, G., Rusciano, G., Pesce, G., Dochshanov, A., Sasso, A. *Nanoscale* **2015**, *7*, 8593–8606.

65. Jin, Q., Li, M., Polat, B., Paidi, S. K., Dai, A., Zhang, A., Pagaduan, J. V., Barman, I., Gracias, D. H. *Angew. Chem. Int. Ed.* **2017**, *56*, 3822–3826.

66. Hu, Q., Tay, L.-L., Noestheden, M., Pezacki, J. P. *J. Am. Chem. Soc.* **2007**, *129*, 14–15.

67. Stanca, S. E., Matthäus, C., Neugebauer, U., Nietzsche, S., Fritzsche, W., Dellith, J., Heintzmann, R., Weber, K., Deckert, V., Krafft, C. *Nanomed. Nanotechnol. Biol. Med.* **2015**, *11*, 1831–1839.

68. Xiao, M., Lin, L., Li, Z., Liu, J., Hong, S., Li, Y., Zheng, M., Duan, X., Chen, X. *Chem. An Asian J.* **2014**, *9*, 2040–2044.

69. Fogarty, S. W., Patel, I. I., Martin, F. L., Fullwood, N. J. *PLoS One* **2014**, *9*, e106283.

70. Hodges, M. D., Kelly, J. G., Bentley, A. J., Fogarty, S., Patel, I. I., Martin, F. L., Fullwood, N. J. *ACS Nano* **2011**, *5*, 9535–9541.

71. Böhme, R., Richter, M., Cialla, D., Rösch, P., Deckert, V., Popp, J. *J. Raman Spectrosc.* **2009**, *40*, 1452–1457.

72. Böhme, R., Cialla, D., Richter, M., Rösch, P., Popp, J., Deckert, V. *J. Biophoton.* **2010**, *3*, 455–461.

73. Opilik, L., Bauer, T., Schmid, T., Stadler, J., Zenobi, R. *PCCP* **2011**, *13*, 9978–9981.

74. Richter, M., Hedegaard, M., Deckert-Gaudig, T., Lampen, P., Deckert, V. *Small* **2011**, *7*, 209–214.

75. Wang, H., Schultz, Z. D. *ChemPhysChem* **2014**, *15*, 3944–3949.

76. Xiao, L., Wang, H., Schultz, Z. D. *Anal. Chem.* **2016**, *88*, 6547–6553.

4 Scanning Angle Interference Microscopy (SAIM)
Theory, Acquisition, Analysis and Biological Applications

Cristina Bertocchi, Timothy J. Rudge,
and Andrea Ravasio

CONTENTS

4.1   INTRODUCTION

Biological molecules are organized into structural and signaling ensembles that work together as nanomachines to facilitate cellular function.[1] A fundamental requirement in biomedical research is to establish a fully integrated understanding of how such nanoscale building blocks are assembled into supramolecular complexes and how they work together in health and disease, at the molecular level. In other words, it is essential to map out the "blueprints" of how the core components of such complex machines are organized. Such knowledge will allow for recognition of the onset and prediction of the effect of disease and formulate a proper and effective means to prevent and treat disease.

  Due to the nanometer size of proteins, investigation of the structure and dynamics of such biological complexes requires spatial resolution at the nanometer scale (protein scale). The spatial resolution of conventional light microscopy has long been hampered due to the diffraction of light[2,3] through lenses and circular apertures. The spatial resolution of fluorescence microscopy was pioneered in 1873 by Ernst Karl Abbe,[2,3] who stated that due to the wave-like behavior of light, the smallest resolvable distance between two objects using a conventional microscope may never be smaller than approximately half of the wavelength of the light used to image the specimen. This poses a limit to resolve two objects as separated ones when they are at a distance less than what is now known as the Rayleigh (R) resolution criteria[4]:

$$R(x, y) = \frac{0.61\lambda}{NA} \tag{4.1}$$

$$R(z) = \frac{2n\lambda}{(NA)^2}, \tag{4.2}$$

where λ is the wavelength of light, n is the refractive index of the specimen, and NA is the numerical aperture of the objective. Thus, with 1.65 being the largest numerical aperture available (1.49 more common), the limit of resolution prevents objects smaller than approximately 200–250 nm in the lateral (x, y) dimensions and ~600–850 nm in the axial (z) dimension from being visualized as anything but blurred spots. This is due to the overlap of their respective point spread functions (PSFs) as spread by the microscope imaging system. This limit is quite stringent for imaging subcellular structures such as protein complexes. As a result, beginning in the late 1980s, such limitations served as the motivation for the development of super-resolve fluorescence microscopy (also known as optical nanoscopy).[5–8] Such methods, already reviewed elsewhere,[9–11] tackle nanoscale imaging through different strategies, and resulted in Eric Betzig, Stefan Hell and William E. Moerner being awarded Nobel Prize in Chemistry in 2014.[12]

From the first detection of a single molecule in 1989[12] and during almost three decades that led to current technical advances in superresolution fluorescence imaging, it has become possible to uncover new features with unprecedented details and to probe previously unattainable components of the intricate nanoscale architecture, dynamics and functions of several cellular complexes. Superresolution methodologies continue to evolve, with new improvements that allow tailoring the available techniques to particular needs and applications. Such developments have enabled biologists to access nanoscale organization, thus elucidating the nanoscopic structures in cellular membranes. Single molecule localization (SML) techniques such as STORM (stochastic optical reconstruction microscopy)[6] and PALM (photo-activated localization microscopy),[7] using photoconvertible or photoactivatable fluorescent proteins, can resolve ~20 nm in the lateral (xy) plane. Nevertheless, attaining similar resolution in the axial (z) direction is more challenging, with a maximum limit at ~50 nm.[13] An improvement to such techniques is the combination of PALM with single-photon simultaneous multiphase interferometry that provides sub-20 nm 3D protein localization, a method known as iPALM (interferometric Photo-Activated Localization Microscopy).[14] Although it provides exceptional sensitivity and specificity for detecting and visualizing molecules in cells, the complexity of the optical system for iPALM and its impracticability for non-specialists limit its widespread use. Furthermore, SML methods generally have poor temporal resolution[15] and are typically performed on fixed cells due to the requirement for long duration for sequential image acquisition to faithfully reconstruct the sample at high resolution.

An alternative method, capable of nanometer resolution on the z plane, but much less complex and more suitable to live imaging is the structured-illumination microscopy (SIM) technique.[16,17] Spatial resolution in SIM is increased by replacing the uniform illumination in a conventional widefield microscope with a series of excitation light patterns containing fine details. The interference between such an illumination pattern with normally unobservable high spatial frequency information present in the sample results in coarsely patterned Moiré (fringe) images. The recorded images are processed to extract information and produce a reconstruction that has improved lateral resolution (2D SIM) or both lateral and axial resolution (3D SIM) with approximately twice the normal resolution.

In the same family of structure-illumination techniques we developed surface-generated fluorescence interference contrast methods.[18–22] The idea behind these approaches relies on creating patterns of varying intensity by constructive and destructive interference of the direct and reflected light at the surface of an oxidized silicon wafer. Unlike STORM and PALM, these methods usually acquire images from large numbers of fluorophores rather than single molecules, and thus do not result in *true* resolving techniques. In general, with these methods samples are prepared above a transparent spacer (silicon oxide) on top of a reflective surface (silicon wafer) (Figure 4.1c and e). A standing wave with specific interference patterns is generated by the interference of the incident light and its own reflection[23] and results in an excitation intensity dependent on the distance (height) of the fluorophore

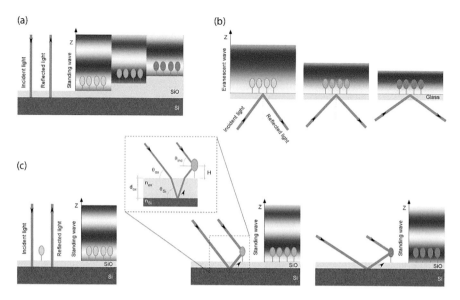

FIGURE 4.1 Principles of fluorescence interference contrast microscopy methods.[18–22] (a) Principles of FLIC.[18–20] In FLIC, silicon oxide layers of different thickness are used as spacers between the sample and the reflective silicon surface. Different interference patterns are created by interference between incident and reflected light from the Si surface depending on the thickness of the oxide layer. The fluorescence intensity depends on the position of the dye with respect to the nodes and antinodes of the standing wave. (b) Principles of VA-TIRFM.[29] In VA-TIRFM, the characteristic 1/e decay of the evanescent field is modulated by varying the incidence angle at the TIR interface. Fluorophores in close proximity to a cell–substrate interface are excited by an evanescent wave with variable penetration depth and localized with high (nanometer) axial z-resolution. (c) Principles of SAIM[22] (and VIA-FLIC[21]). In SAIM, a silicon wafer with a silicon oxide layer of only one defined thickness is used as a spacer. The reflection of coherent excitation light interferes with the incoming beam and generates a standing wave pattern, resulting in fluorescence intensity that varies with distance to the surface. The standing wave pattern varies as a function of the incidence angle of the light. The schematic described for SAIM apply also to VIA-FLIC. The only difference is that instead of varying the standing wave pattern by modifying the incidence angle of the light, in VIA-FLIC, the change is obtained by the replacing the aperture diaphragm of an epifluorescence microscope with annular photomasks with different dimensions. These photomasks will create a hollow cone of excitation with a specific narrow range of angle of incidence depending on the different radial positions in the diaphragm plane. In all these schematic diagrams (a–c), low intensity excitation light due to interference is denoted by dark blue, shading to high excitation intensity in light blue. In (a–c) A fluorophore at a height corresponding with constructive interference will register higher fluorescence intensity (bright green fluorophore) than a fluorophore at a height corresponding with destructive interference (dark green fluorophore).

above the reflective surface. The excitation intensity pattern directly modulates the rate of absorption of a fluorescent object as a function of its z-position. Since the fluorescence emission intensity is dependent on the objects absorption rate, knowing the excitation pattern allows extraction of distance information. By varying the pattern of excitation, different intensity profiles are obtained, and the fluorescence intensity can be related to the object's distance from the interface. After acquiring images, the measured fluorescence intensity profiles are fitted pixel by pixel to a theoretical optical model to estimate the average height of each x-y element at nm resolution (with 5–10 nm precision). In the context of surface-generated structured illumination techniques, the reconstruction resolution is the minimum z displacement at which two fluorescent objects can be separately identified by numerical processing of images obtained with different excitation intensity functions. The level of attainable resolution depends strongly on the maximal gradient of the excitation intensity and the signal/noise ratio of the images. A great advantage of these methods is that they utilize relatively simple optics available on widely available total internal reflection fluorescence microscopes, offering potential for widespread use in cell biology.[22,24–28]

4.2 THEORY

4.2.1 Fluorescence Interference Contrast (FLIC) Microscopy[18–20]

Developed by Fromherz and co-workers,[18–20] FLIC uses surface-generated structured illumination to interrogate the vertical positions of nanometer-sized objects (Figure 4.1a). This is possible by introducing a reflective surface (a mirror) behind the sample to create axially varying illumination. In practical terms, silicon oxide layers of different thickness are used as spacers between the sample and the reflective silicon surface. The interference pattern generated when light is reflected from these surfaces interacts in predictable patterns with fluorophores situated on them (Figure 4.1a). Indeed, the fluorescence intensity depends on the position of the fluorophore with respect to the standing waves of light in front of the reflecting silicon surface. For example, a fluorophore at a height corresponding to constructive interference will register higher fluorescence intensity than a fluorophore at a height corresponding to destructive interference. By relating the interference pattern to fluorescence intensity, it is possible to calculate fluorophore's absolute height above the reflective surface. Note that FLIC chips/wafers with different oxide layers can be fabricated using relatively straightforward lithographic methods,[20] but the different oxide layers have micron-sized lateral x-y dimensions and generally only objects uniformly fluorescently labeled that are large enough to span over the multiple layers along the optical axis can be imaged.

4.2.2 Variable-Angle Total Internal Reflection Fluorescence Microscopy (VA-TIRFM)[29]

In total internal reflection fluorescence microscopy (TIRFM),[30] an evanescent field of excitation light is generated when a propagated wave strikes the boundary between two media with different refractive indices at an angle larger than a particular critical angle with respect to the normal to the surface (Figure 4.1b). The evanescent field decays exponentially from the total internal reflection (TIR) interface with a characteristic 1/e distance. Only fluorophores within approximately 200 nm of the TIR interface are excited to an appreciable extent, the objects outside of the evanescent field are cut out and are not excited. In VA-TIRFM, also referred to as multiple angle-TIRFM, the characteristic 1/e distance of the evanescent field is modulated by varying the incidence angle at the TIR interface (Figure 4.1b). The incident light is delivered via a monomode fiber and focused onto a sample that is optically coupled to a hemi-cylindrical glass prism, under different angles of total internal reflection. Thus, fluorophores in close proximity to a cell–substrate interface are excited by an evanescent wave with variable penetration depth and localized with high (nanometer) axial z-resolution. Although VA-TIRFM could be used to image isolated objects, its applicability has been hampered by its experimental complexity, especially due to the need for a through-prism TIRFM geometry, to achieve the range of incidence angles required.

4.2.3 VARIABLE INCIDENCE ANGLE FLIC (VIA-FLIC)[21]

The VIA-FLIC technique, developed by Ajo Franklin and co-workers[21] is a combination of FLIC and VA-TIRFM. Like FLIC, it uses interference from a very flat silicon mirror to create structured illumination but instead of using SiO_2 (silicon oxide) steps, the incidence angle of excitation light is modified by the introduction of custom-fabricated annular photomasks with different radii. These photomasks are placed in the aperture diaphragm plane of the microscope to create hollow cones of excitation light. Simply modulating the incidence angle alters the periodicity of the interference and hence the intensity of detected fluorescence at different heights above the surface. VIA-FLIC has the potential to visualize the rapid movements and spatial organization of proteins. It is worth considering though that many data points taken with different photomasks are necessary to obtain an accurate quantitative fit, affecting the time resolution of the method.

4.2.4 SCANNING ANGLE INTERFERENCE MICROSCOPY (SAIM)[22]

SAIM was described for the first time by Paszek and colleagues[22] as an improvement of the theory and methodology of the above described surface-generated structured illumination microscopy techniques (Figure 4.1c). In SAIM, a silicon wafer with a silicon oxide layer of only one defined thickness is used as a spacer, thereby simplifying substrate preparation. The reflection of coherent excitation light interferes with the incoming beam and generates a standing wave pattern, resulting in fluorescence intensity that varies with distance from the surface. The incidence angle of the light is varied, modifying the standing wave pattern, and a series of images at different angles between the center and the maximum incident angle is recorded. By fitting the raw interference images to a mathematical model that describes how excitation intensity varies as a function of height and incidence angle of the excitation laser, it is possible to reconstruct the z-position of fluorescent molecules with 10 nm or better precision.

The basic hardware requirements for SAIM are a fluorescence microscope, a coherent excitation source (i.e., a laser), motorized optics to focus the source at defined locations on the back aperture of the microscope objective, and a camera detector. These requirements are satisfied by the current generation of motorized total internal reflection fluorescence (TIRF) microscope systems offered by major manufacturers. Although conducted on a TIRF microscope, SAIM is not actually performed with laser incident angles that exceed the critical angle for total internal reflection. Rather, the TIRF illuminator provides a simple way for controlling the position of the excitation laser beam on the back aperture of the objective and hence the incident angle of the excitation beam out of the front lens of the objective.

SAIM is not a "true" superresolution technique as it cannot resolve the position of individual fluorophores within a diffraction-limited spot but it is capable of reporting the average vertical position of fluorophores within a diffraction-limited volume. Furthermore, the calibration, acquisition, and analysis for SAIM are relatively simple, making it a suitable method to dynamically image nanotopographical features of biological structures of nanometer thickness (i.e., structures whose vertical thickness is ≤ 150 nm, such as the plasma membrane, cytoskeleton, membrane associated complexes, and intracellular vesicles).

4.3 OPTICAL PRINCIPLES

Scanning angle interference microscopy is based on the interference of coherent light incident on the sample, with the same light reflected from the silicon surface (see Figure 4.1). This interference occurs due to the differing path lengths of light directly illuminating the sample, at some height above the silicon layer, and light reflected from the silicon layer, passing through the silicon oxide to finally hit the sample. This sets up a standing wave illumination pattern, which varies in intensity (or electric field strength) with height above the silicon surface and depends on the angle of incident light as will be outlined below. The surface-generated standing wave as a function of incidence angle (θ_{inc}) and fluorophore z-position (nm) is indicated in Figure 4.2a.

It can be shown[18,22] that the electric field strength of plane wave incident light on a flat surface, where the light is polarized perpendicular to the surface, is approximated by

$$F \approx 1 + r^{TE} e^{i\phi(H)}, \qquad (4.3)$$

where H is height above the surface, r^{TE} is the transverse electric component of the effective Fresnel coefficient of reflection of the layered silicon-silicon oxide surface, and $\phi(H)$ the phase difference between the direct and reflected light at height H above the substrate

$$\phi(H) = \frac{4\pi}{\lambda} \left(n_b H \cos\theta_{inc} \right), \qquad (4.4)$$

for light with wavelength λ and incident angle on the sample θ_{inc}, which has refractive index n_b.

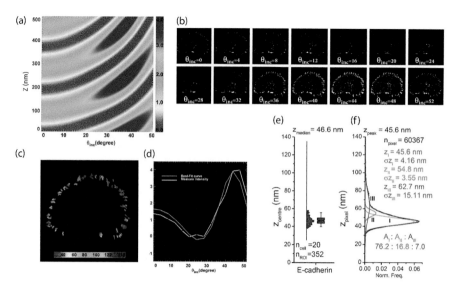

FIGURE 4.2 Nanoscale fluorescence imaging methods. (Suppl. Figure 4.2 from[25] with minor modifications.) (a) Surface-generated standing wave as a function of incidence angle: intensity of the fluorescence excitation field as a function of incidence angle (θ inc) and fluorophore z-position (nm). (b) Montage of raw data from E-cadherin-EGFP expressed in MDCK cell on Ecadherin biomimetic substrates, showing the variation of fluorescence intensity with θ inc (degree). (c) Topographic z-position map of the same fluorophore as in (b). Z-position is calculated pixel-by-pixel by least-square fitting of the measured angle-dependence curve (white) to theoretical model (green), as shown in (d). (e) Notched box plots and histograms for z-position calculated from the median of each adhesion ROI: first and third quartiles, median, and confidence intervals; whiskers, 5th and 95th percentiles. Histogram bin size, 1 nm. The median of this distribution is the representative protein z-position, zcenter. $n = 352$ adhesions (nROI) pooled from 20 cells (ncell). (f) Profile of E-cadherin distribution along the z-axis. Normalized histogram (black, 1-nm bin) of pixel z-position. Also shown are the decomposition into 3 Gaussian functions along with the Gaussian centers (zI, zII, zIII), widths (σzI, σzII, σzIII), and relative amplitude (AI, AII, AIII). Peak z-position of the distribution (zpeak) and number of pixels analyzed (npixel) indicated. (Legend for the figure has been included as in the original paper[25] with minor modifications.)

The complex Fresnel coefficient r^{TE} is typically calculated using the transfer matrix method. Consider the case of a silicon surface with a single oxide layer of thickness d_{ox} with refractive index n_{ox} (Figure 4.1e). The angles of incidence ($\theta_{inc}, \theta_{ox}, \theta_{Si}$) in each medium (sample, oxide and silicon, respectively) are given by Snell's law:

$$\frac{\sin\theta 1}{\sin\theta 2} = \frac{v1}{v2} = \frac{n1}{n2}, \quad (4.5)$$

that describes the refraction of light at the interface between two media of different refractive indices (n), with $n2 > n1$. The characteristic transfer matrix is then given by:

$$M_{TE} = \begin{pmatrix} m_{11}^{TE} & m_{12}^{TE} \\ m_{21}^{TE} & m_{22}^{TE} \end{pmatrix} = \begin{pmatrix} \cos\left(k_{ox}d_{ox}\cos\theta_{ox}\right) & \dfrac{-i}{p_1}\sin\left(k_{ox}d_{ox}\cos\theta_{ox}\right) \\ -ip_1\sin\left(k_{ox}d_{ox}\cos\theta_{ox}\right) & \cos\left(k_{ox}d_{ox}\cos\theta_{ox}\right) \end{pmatrix}, \quad (4.6)$$

where $k_{ox} = \frac{2\pi n_{ox}}{\lambda}$ is the wavenumber in the oxide layer and $p_1 = n_{ox}\cos\theta_{ox}$. The reflection coefficient is then given by:

$$r^{TE} = \frac{\left(m_{11}^{TE} + m_{12}^{TE}p_0\right)p_2 + \left(m_{21}^{TE} - m_{22}^{TE}p_0\right)}{\left(m_{11}^{TE} + m_{12}^{TE}p_0\right)p_2 + \left(m_{21}^{TE} + m_{22}^{TE}p_0\right)}, \quad (4.7)$$

with $p_0 = n_{Si}\cos\theta_{Si}$, $p_2 = n_b\cos\theta_{inc}$.

After calculation of the transfer matrix and the resultant Fresnel coefficient from the geometry of the experimental setup (Figure 4.1), we are left with a nonlinear model for the intensity of incident light at height H, which depends on the angle of incidence of illuminating light. From equation (4.3),

$$I = A\left|1 + r^{TE}e^{i\phi(H)}\right|^2 + B \quad (4.8)$$

It is assumed that the emitted light intensity is proportional to the illumination intensity, and hence the image pixel intensity at any given scanning angle θ_{inc}. Here we have introduced a constant of scaling (A) to account for laser intensity, fluorophore density, detection efficiency and optical effects of self-interference. The offset parameter B simply accounts for background fluorescence in the sample.

In SAIM the angle of incident light is scanned in a range up to the critical angle for total internal reflection, and a series of images are recorded. For each pixel (or the average of some image region), the intensity varies with angle according to equation (4.8). Figure 4.2d shows a typical intensity profile. Nonlinear least squares or other optimization methods can then be used to estimate the parameters A, B, and H of such profiles by fitting to equation (4.8).

4.4   EXPERIMENTAL METHODS AND INSTRUMENTATION

4.4.1   PREPARATION OF REFLECTIVE SUBSTRATES

Commercially available silicon wafers with nanometer-sized thermal SiO_2 layers (i.e., Bonda Technology, Wafer World, etc.) can be used for sample preparation in SAIM. Alternatively, silicon oxide wafers could be prepared according to the method described in Paszek et al.[22] In either case, it is critical to determine the precise thermal oxide thickness of each batch of silicon wafers with nanometer-precision because this could affect the read-out of the true height of a sample by SAIM. Measurements of the oxide thickness could be obtained using a UV-visible variable angle spectroscopic ellipsometer (UV-VIS-VASE). The variations in oxide thickness across the wafers should be typically less than 1.5 nm. Oxide spacer thickness can also be estimated using SAIM by imaging a monolayer of surface-bound fluorescent dyes (such DiO, DiI, and DiD in supported lipid bilayer[31]) of defined thickness and fitting theoretical predictions for different oxide heights to the data to derive the known height of the sample. To maximize interference contrast, N-type (100)-orientation silicon wafers with ≥500 nm of silicon oxide are recommended,[22,24,25] as it provides the ideal spacing between the sample and the reflective surface however different thicknesses have been successfully used. Each wafer is cut into 1 × 1 cm chips using a diamond-tip pen, followed by sequential cleaning using acetone and 1M KOH with sonication for 20 minutes each, and four washes in dH$_2$O in between. The wafers are then chemically activated (functionalized) by silane to enable conjugation of matrix proteins for cell adhesion. The type of silane must be chosen and optimized according to the silane chemistry[32] and the functional group of the protein to which it will be linked in order to allow for a specific type of cell adhesion. For fibronectin or collagen type 1 conjugation,[22,24] wafers are incubated for 1 hour with the silane of choice would be either (3-aminopropyl) trimethoxysilane (APS) at 0.5% in water, or (3-aminopropyl) triethoxy silane (APTES) at 0.5% in PBS. Wafers would then be sonicated five times in water for 5 minutes to remove excess APS or APTES, followed by incubation in 0.5% glutaraldehyde in PBS for 1 hour, and then sonication five times each in water each for 5 min. For IgG conjugation (for the preparation of cadherin biomimetic substrates[25]), (3-Glycidoxypropyl) methyldimethoxysilane can be used at a concentration of 0.045% in 100% Ethanol for 1 hour, cured at 110°C for 1 hour and rinsed in 70% ethanol first and then dH$_2$O. Silanized wafers are air-dried by nitrogen gas and can be stored for prolonged periods of time until use, or immediately conjugated to the protein of interest diluted in a proper buffer (attention must be paid to the correct pH) at proper temperature and incubation time (see Section 4.4.4).

4.4.2 PREPARATION OF REFLECTIVE SUBSTRATES WITH ADSORBED NANOBEADS

Precision of measurements by SAIM can be assessed with the use of nanobeads of different diameter. Nanobeads are adsorbed onto silicon substrates with the same oxide spacer as for the measurements (~500 nm height, in our experimental setup[25]) with a simple protocol according to the method described in Carbone et al.[31] Briefly, 40- or 100-nm carboxylate-modified yellow-green, orange, or red fluorescent spheres are diluted in 70% ethanol, added to the wafers, and dried in a vacuum desiccator. The wafers are then washed vigorously with water, air-dried, and stored at room temperature. Angle dependent changes in nanobead fluorescence intensity can be fitted to the optical model to obtain the axial (z) position of the bead center. These measurements should be acquired for nanobeads with nominal radii of 20-, 50- or 100-nm, and fluorescence excitation wavelengths of 488 or 561 nm. To further validate SAIM height measurements for the nanobeads, it is advisable to compare them to measurements obtained by EM as suggested by Carbone et al.[31]

4.4.3 PREPARATION OF FLUORESCENT REFERENCE SLIDES

Variations of intensity in laser illumination in wide-field microscopy could lead to interference fringes. To quantify such variation in intensity, reference slides with a monolayer of fluorescent dye can be prepared using silane conjugates of fluorescein and rhodamine B. The conjugates can be synthesized according to the protocol described in Paszek et al.[22] Briefly, 1 mg of fluorescein isocyanate or 1.37 mg of rhodamine B isocyanate is made to react with 9.25 mg of APS in 1.25 mL anhydrous ethanol, under nitrogen gas at ~21°C for 2 hours with constant stirring. Immediately following the reaction, the conjugates are centrifuged at 20,000 g for 10 min to pellet unwanted aggregates, and then diluted 10-fold in absolute ethanol. Clean glass-bottom dishes (#1.5 glass) are incubated with the diluted silane conjugates for 30 minutes at ~21°C, rinsed with water and stored in water until imaging.

4.4.4 PREPARATION OF THE IMAGING SAMPLE

Chemically activated (silanized) wafers are sterilized under a UV lamp for a few minutes before use. For protein conjugation, as seen for the silane (see Section 4.4.1), the method of conjugation will vary with respect to the protein to be conjugated.

For fibronectin conjugation, specific for integrin binding (i.e., to study focal adhesion nanoscale structure and dynamics), the silanized wafers are incubated in PBS with plasma fibronectin [10 µg/mL] in a humidity chamber for 40 minutes at 37°C or overnight at 4°C.[22,24]

For E- or N-cadherin conjugation,[25] specific for E- or N-cadherin homophilic binding (i.e., to study adherins junction nanoscale structure and dynamics), the silanized wafers are incubated overnight at 4°C in a humidity chamber with anti-F_c fragment specific antibody for E- or N-cadherin [1 µg/cm^2] in 0.1 M pH 8 borate buffer (pH 8.3). The next day, wafers are washed with PBS and neutralized with 2-(2-Aminoethoxy) ethanol for 1 hour at RT. After rinsing with PBS, the substrates are incubated with E- or N-cadherin-Fc chimeric protein for 2 hours at RT [1 µg/cm^2], rinsed with PBS (with Ca^{2+} and Mg^{2+}, to avoid any collapse of cadherin-cadherin binding structure). Also, attention must be paid not to expose cadherin substrates to air and always maintain them wet. After protein conjugation, wafers are first washed in PBS and then blocked for 20 minutes to 1 hour with 0.2% Pluronic acid to allow for complete inhibition of non-specific binding. After washing with PBS, cells can be plated on the protein-conjugated silicon wafers.

4.4.5 FLUOROPHORES FOR SAIM

SAIM is feasible for fixed samples as well as live-cell imaging. To avoid loss of accuracy we do recommend special attention to the choice of fluorophores. Although SAIM does not require special fluorophores, the fluorophores should have high photostability to minimize photobleaching during the imaging scanning sequence and should be bright with high quantum yield to provide a good signal-to-noise ratio. Among the fluorescent proteins and synthetic dyes compatible with SAIM, genetically encoded fluorescent proteins have the principal advantage of being small and suitable for live-cell imaging and are capable of achieving maximal labeling specificity, removing any possible problems associated with nonspecific labeling. Furthermore, they do not require fixation or permeabilization procedures that could affect cellular nanostructure. Some fluorescent proteins successfully used in SAIM include green fluorescent proteins such as EGFP and mEmerald, red fluorescent protein mCherry,[33] and photoconverted tdEOS[25] that has excellent brightness and photostability. In addition, chromobodies (generated by the fusion of a fluorescent protein to a nanobody[34]), a novel species of extremely small antibodies that are endogenously synthesized within cultured cells, could possibly be used to prepare samples for SAIM.

4.4.6 MICROSCOPE AND INSTRUMENTATION

SAIM imaging can be performed on an inverted microscope equipped with a motorized objective-type TIRF illuminator and with lasers suitable for excitation of the fluorophores that have previously been selected (fused or conjugated to molecules of interest). The excitation source should be linearly polarized perpendicular to the plane of incidence along which the laser light emanates from the objective (to maintain s-polarization). A glass linear polarizing filter is the only hardware modification that is required to achieve this linear polarization and perform SAIM on a commercially available motorized TIRF system. The polarizer can be placed in the filter cube housing the dichroic mirror. DuFort and colleagues found that the slight ellipticity in polarization that is introduced by optics downstream of the polarizer does not significantly affect imaging with SAIM.[35] For the best overall performance, it is important to consider the selection of the microscope, the objective, and the camera. Since SAIM requires imaging through a relatively thick aqueous buffer, optimal performance and maximum resolution are achieved with a high NA water immersion objective coupled with a detector that has appropriate pixel sizes to satisfy the Nyquist criterion.[36] For example, a possible pairing could be a 60x NA 1.49 ApoTIRF objective lens with a low noise, scientific complementary metal oxide semiconductor (sCMOS) camera with small pixel sizes (6.5 × 6.5 μm) and a global shutter mode.[25,35] However, other objective and camera configurations can provide suitable performance, depending on the imaging criteria.

Since the fluorescent sample is imaged with a coherent light source in SAIM, the microscope optics should be as clean as possible to minimize uneven illumination due to laser fringing that could interfere with the interference pattern. As a rule of thumb, filters, dichroic mirrors and lenses should be kept spotless; thus periodic cleaning with a jet of air or more stringent methods are strongly recommended.

4.4.7 SYSTEM CALIBRATION

Prior to acquisition of the imaging scanning sequence for reconstruction, a calibration of the system should always be performed. In particular, it is necessary to calibrate the position of the motor of the TIRF illuminator for the actual refractive index corrected angle of excitation light for each wavelength used. This can easily be achieved by manually varying the motor position of the TIRF illuminator and measuring the position of the laser angle out of the objective, starting from the angle 0° (TIRF center) adjusting perpendicular vertically from the objective front lens, until below the critical angle. To ease the calibration, a printed grid adhered to a flat support and placed on the microscope stage can be used to precisely calculate the angle out of the objective (with basic trigonometric relations). Calibration is typically performed in air using a clean 27-mm glass bottom dish (Iwaki) placed on the microscope stage holder. Subsequently, the angles in air must be corrected for the corresponding angles in aqueous samples, according to Snell's Law (equation 4.5).

The tabulated motor positions for the angle 0° to +56° (below the critical angle with 4° increments) obtained by application of Snell's law (equation 4.5) and corresponding to the desired laser incidence angles for image acquisition can then be programmed as a multidimensional acquisition into the acquisition software (e.g., NIS-elements, MetaMorph, μManager, etc.) to facilitate automated and rapid data acquisition.

To ensure correct calibration of the system and verify the accuracy of the measurements, it is recommended to run a test with the adsorbed nanobead slides previously prepared (see Section 4.4.2). If properly configured, SAIM will report the centroid height (z_{center}) of the beads above the surface (i.e., if using 100 nm nanobeads, the system will indicate a centroid of ~50–60 nm).

As convenient alternative to the manual calibration, Carbone and colleagues have developed a method for automated angle calibration based on a μManager controlled Arduino device, freely available as an open source tool online (Development, under the Berkeley Software Distribution (BSD) license, and is hosted by GitHub at https://github.com/kcarbone/SAIM_calibration).[31]

4.4.8 ACQUISITION

For imaging, the silanized silicon chip is placed facing the objective with the sample (and thermal oxide) side on the glass surface of a 27-mm glass bottom dish (Iwaki) filled with aqueous buffer (i.e., PBS, PHEM, phenol red-free growth medium, etc.). A small thumbscrew (of ~1.75 g) is placed on top of the chip to keep the wafer sample surface immobilized and flat against the glass bottom (to maintain neutral buoyancy). For live-cell imaging, the microscope chamber must be kept at 37°C and supplied with CO_2 (unless using a medium buffered for pH, i.e. supplemented with HEPES). SAIM requires the acquisition of a sequence of images at consecutive calibrated laser angles of incidence (raw images in Figure 4.2b). Desired laser incidence angles for image acquisition (each corresponding to a unique TIRF motor position) can be programmed as a multidimensional acquisition into acquisition software (including, but not limited to, NIS-elements, MetaMorph and μManager) to facilitate automated and rapid data acquisition. Typically, for a wafer with 500 nm oxide height, the scanning sequence can start from the left −52° till 0° and then to the right +52° with 4° increments or only in one direction to avoid redundancies. Nevertheless, the optimal range of angles and step sizes should be optimized according to the expected sample height above the substrate. This requires a trade-off between rate and range, because it should permit full sampling of the periodic intensity profile (more frequent inversions in fluorescence intensity can be expected for higher structures). At the completion of the laser scanning sequence acquisition, the series of raw images is saved as. ND2 file, ready for conversion, processing, analysis and reconstruction. It is important to note the excitation wavelength (since the refractive index of buffer (H_2O), Si, and SiO_2 are wavelength-dependent), pixel size, magnification, the incidence angles and oxide thickness for each acquisition as this information will be necessary in the analysis phase. Also, to allow for comparison between measurements, all these parameters should be maintained exactly the same across different acquisitions.

As mentioned above, SAIM is suitable for imaging with fixed samples or live cell imaging, in one single position or in multiple stage positions; in this last case, the use of hardware autofocus (e.g., Nikon Perfect Focus System (PFS)) is critical. The distance between imaging positions should not be too large to minimize changes in the position of the wafers due to its movement during travel. Minimal X, Y shifts can always be compensated or adjusted with the analysis software or with an image J plugin.

4.4.9 ANALYSIS

For the analysis, images saved as ND2 must be exported as Multi-Page TIFF files (TIFF stack, 16-bit, single channel). Such a file is a stack of raw images that allows the reconstruction of the nanoscale height by analyzing intensity profiles on a pixel-by-pixel basis across the image stack.

The software we have used in the laboratory of Prof. Kanchanawong for the analysis of SAIM data is called Zmap3.[24,25] This program is a compiled IDL runtime application (IDL can be downloaded from Exelis VIS for free, after registration), based in part on the MATLAB source code kindly provided by C. Ajo-Franklin, Lawrence Berkeley National Laboratory, Berkeley, CA.[21]

TIFF images are opened in Zmap3, all relevant parameters (excitation wavelength, pixel size, magnification, thickness of SiO_2 layer above the Si layer, refractive index of SiO_2 layer, maximum incidence angle and interval of incidence angle) associated to the image acquisition for that particular image (or stack of images) are set, and the regions of interest (ROI) to be binarized and analyzed are defined and saved as *zmap.sav files.

After we obtain a binary image to mask the ROI of the object of interest (either by simple thresholding or by Otsu thresholding using background subtracted images), we can calculate the angular dependence of the intensity of the object at a specified Z height. Topographic height (z) and other fit parameters are determined by fitting the theoretical angle-dependence curve to experimental angle-dependence intensity by the Levenberg–Marquardt or similar nonlinear least squares methods for each pixel in the ROI. Topographical Z values are displayed using color to encode z-position above the Si surface as a Zmap of the region of interest and a plot of the angular dependence at different Z heights together with the mathematical fit (Figure 4.2b and c). The fit should follow closely the experimental data. This could be determined visually and with the help of the uncertainty value expressed as sigma (the fit can be considered good if the uncertainty value is ≤7 nm). Sigma Z roughly corresponds to "resolution" and it should be as low as possible. Topographical Z values obtained from several measurements (n) can be represented as notched box plots and associated histograms for z-position calculated from the median of each adhesion ROI, as indicated in Figure 4.2e, or as a distribution profile along the z-axis, as exemplified in Figure 4.2f.

Carbone and colleagues[31] in a recent publication described a straightforward pipeline for SAIM analysis available under the Berkeley Software Distribution (BSD) license (https://github.com/nicost/saimAnalysis/).

4.5 MULTICOLOR SAIM

An additional advantage of SAIM is the ability to image multiple color channels. Measurements are performed sequentially, using excitations corresponding to different fluorophores fused to the proteins of interest, and following the same procedure as indicated in the acquisition Section 4.4.8. Analysis is performed as previously indicated in the analysis Section 4.4.9. The scatter graph of z-position for each pixel of two different fluorophores are plotted, and a linear regression calculated using the software OriginPro. This option is very convenient when we look for orientation of specific proteins or at potential interacting partners in a complex.

4.6 SAIM ON SUBSTRATES WITH DIFFERENT RIGIDITIES

The advancement in super-resolution imaging techniques has allowed researchers to probe the organization and localization of proteins within the cell with nanoscale precision. One potential drawback is that current technologies rely on the use of rigid substrates, such as glass and silicon wafers. However, with the advent of mechanobiology,[37,38] the importance of mechanical properties of the extracellular matrix in regulating many aspects of cellular behavior and signaling has become more and more accepted.[39] The ability to mechanically tune the cell substrate *in vitro* and the ability to perform super-resolution microscopic imaging at the nanoscale, can thus significantly enhance the physiological relevance of a study. Guanqing Ou and co-workers[40] have developed a system that allows SAIM on a mechanically tunable substrate and have shown its application for nanoscale imaging of subcellular components on mechanically tunable substrates.[40] Briefly, mixed and degassed silicon gel solution (Corning 52–276 Parts A and B at ratios corresponding to desired stiffness) is deposited on wafer squares using a small pipette tip or dropper before the spin coating program is started. Polymer to crosslinker ratio can be varied to obtain silicone gels of different stiffness, ranging from 250 Pa to 100 kPa, a broad range that spans the physiological range (from 1 kPa in the brain to 25k–100 kPa in muscle and bone[41]). Silicon wafers are then spincoated in thin layers with unpolymerized gel solution at 8000 rpm for 90 seconds. The resulted substrates are baked at 60°C for 3 hours. To conjugate ECM proteins to the substrate surface, the silicon gel-coated wafers are chemically activated by silanization, following the procedure described above (Section 4.4.1).

4.7 APPLICATION OF SAIM TO STUDY BIOLOGICAL MOLECULES

Major advances in cell biology are strongly associated with innovations in microscopy due to its non-invasiveness that permits time-resolved imaging of live cells. In particular, high-resolution microscopy allows extraction of information about the structure–function relationship that is crucial for understanding molecular mechanisms in living cells. In the case of SAIM, axial precision, the ability to image multiple color channels, suitability for live cell imaging due to its dynamic capability, and adaptability to mechanically tuned cell substrates, makes it a useful method for exploration of the molecular mechanisms and topological arrangements underlying dynamic mechanical processes. SAIM is effective for measurement of distinct structures with a thickness <150 nm. This feature makes it perfectly suited for studying topography of plasma membranes and associated protein complexes (e.g., adhesive complexes), intracellular vesicle trafficking (such as exocytosis and endocytosis) associated to the membrane, and organization of cytoskeletal filaments and microtubules. Here we will highlight a few examples of the most recent applications of SAIM to answering biological questions.

4.7.1 Nanotopographical Features of Cell Membrane and Temporal Evolution of the Three-Dimensional Architecture and Nanoscale Dynamics of Focal Adhesion Complexes[22,24,40]

Focal adhesions (FAs) are supramolecular complexes at the interface between cell and extracellular matrix that consist of integrin receptors at the membrane indirectly linked to the actomyosin contractile cytoskeleton through a complicated network of proteins (Figure 4.3). FAs are involved in a plethora of cell biological processes ranging from morphogenesis, immunity, and wound healing; thus, the molecular basis of their nanostructures and the mechanisms regulating force transmission are of key importance to understand their function. In a seminal work, in 2010, Kanchanawong and co-workers imaged the nanoscale protein architecture of FAs using iPALM (Figure 4.3a, b, d). For the first time, they built up the molecular blueprint of this adhesion complex.[42] The paper elegantly demonstrated that proteins in this complex organize into three distinct layers along the vertical axis separating integrin and the actin cytoskeleton. This ~40 nm core is stratified into a signaling layer adjacent to integrin and the membrane, an intermediate force-transduction layer with mechanotransducer elements and an uppermost actin-regulatory layer, with actin and actin regulators. Furthermore, by tagging the two different (n- and c-) termini of talin, a protein of the intermediate layer, they revealed its position and orientation, unraveling its key role in organizing the FA nanostrata, akin to a molecular ruler.

In 2012, Paszek and co-workers presented SAIM,[22] based on a modification of FLIC, allowing the measurement of nanoscale structural details and dynamic behavior in living cells. In order to quantify and validate the performance of this new method, they interrogated the localization of biological structures with nanoscale precision along the optical axis. Specifically, they applied scanning angle interference microscopy in live cells to investigate the temporal evolution of the 3D architecture and nanoscale dynamics of the proteins of the FA complex. They expressed

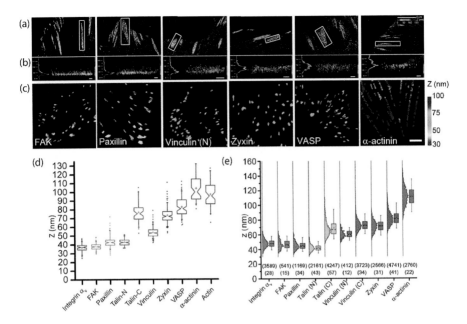

FIGURE 4.3 Protein stratification of the focal adhesion core (this figure is a merge of Figure 1 from[24] and Figures 2 and 4a from[42] with minor modifications). (a) Top view and (b) side view (white boxes, top-view panels) iPALM images and corresponding z histograms and fits of FA proteins in U₂Os cells as indicated in (c): FAK–tdEos; paxillin–tdEos; vinculin–tdEos; zyxin–mEos2; VASP–mEos2; α-actinin–mEos2. Scale bars: 5 μm (a) and 500 nm (b). The vertical distribution of α-actinin is non-Gaussian, so the focal adhesion peak fit is not shown. (c) Topographic maps of FA protein z-positions (nanometers) in HUVECs resulting from SAIM measurements. Color scale bar indicates z-position relative to ECM: 30–100 nm. (d) Peak position (zcenter) of photoactivatable fluorescence proteins in FAs obtained by iPALM measurements. Notched boxes, 1st and 3rd quartiles, median and confidence interval; whiskers, 5th and 95th percentiles; +, means, outliers also shown. (e) Nanoscale stratification of FA protein z-positions obtained by SAIM measurements. Notched boxes and histograms (bin size, 1 nm) for zFA of indicated FA proteins: first and third quartiles, median and confidence intervals; whiskers, 5th and 95th percentiles. Numbers indicated: median zFA (red), number of FA ROIs (black), and number of cells (blue). (Legend for the figure has been included as in the original papers[24,42] with minor modifications.)

talin (the same focal adhesion molecule previously investigated by Kanchanwong and co-workers) fused to two different fluorophores at the N- and C-termini to allow for multicolor SAIM, and they determined its position and orientation, with 37 nm distance between N- and C-terminus and a corresponding 51° molecular orientation with respect to the vertical axis. Furthermore, they measured the position of two adaptor proteins of FA, paxillin and vinculin, fused to two different fluorophores, during cycles of cell retraction in the maturing adhesions of motile cells. Their measurements highlighted a mechanism of downward paxillin movement during cell migration, possibly due to increased mechanical engagement of the cytoskeleton with the adhesion complex.

A further confirmation of the validity and applicability of SAIM to the study of the nanoscale architecture of FAs and their dynamics came from the Kanchanwong's group (Figure 4.3c and e). In their paper,[24] combining superresolution microscopy and engineered talin constructs with alternate lengths, they demonstrated that talin could be considered as the backbone of FAs, serving as the mechanosensitive master coordinator of the FA architecture. Quite noticeably, they cross-verified the topographical Z values obtained by SAIM with iPALM imaging. The precision of measurements with SAIM was comparable to that achieved with iPALM, as shown in the side-by-side graphs of Figure 4.3d and e. Furthermore, the z-positions of FA components such as VASP and F-actin were tested with respect to different engineered talin constructs with alternate lengths, which unveiled a modulation of the physical dimension of FAs along the z-axis mediated by talin.

Overall, SAIM has an excellent performance in the detection of the axial position as compared to the highly sophisticated iPALM technique. Furthermore, it has the advantage of a relatively fast acquisition rate of 1–10 seconds that is suitable for live imaging of FA dynamics, as demonstrated by Paszek and co-worker.[22]

4.7.2 NANOSCALE ARCHITECTURE OF CADHERIN-MEDIATED ADHESIONS[25]

Another excellent example for the application of SAIM is the complex mechano-biological structure of the cadherin-mediated adhesions, also known as adherens junctions (AJs) (Figure 4.4). AJs are considered as a hallmark in evolution of life on earth as their appearance corresponds to the evolution from unicellular organisms organized in colonies to multicellular organisms capable of working as an integrated whole. The main characteristics of this type of adhesions are that they are strong enough to hold the tissue together, but also harbor dynamicity for cell/tissue movements. This is possible thanks to the connection mediated between cadherin and the actomyosin cytoskeleton through a dense plaque of adaptor proteins, known as the cadhesome network.[43] Although the building blocks of this plaque were known, due to the resolution (see Section 4.1) of conventional microscopes and the position of the cadherin-mediated adhesion deep into the monolayer intermixed with other adhesion complexes, the nanoscale organization of the complex had remained a holy grail in cell biology. Making use of SAIM and validating our finding by iPALM technology, we were able to map the position and orientation of key proteins of the cadhesome network, unveiling a striking degree of compartmentalization, highly resembling the structure previously observed in FAs (Figures 4.3 and 4.4a and b). Similar to what is observed in FAs (Figure 4.3), we recognized a stratification of cadherin complex-associated proteins in different layers along the cadherin–F-actin connection with a signaling layer, a force transduction layer (which includes proteins such as vinculin) and actin regulatory layer. Furthermore, with genetic engineering of vinculin mutants and pharmacological treatment, we were able to identify the mechanism that induces the shape-shift conformation of vinculin by both mechanical tension and biochemical signal inputs. In analogy to the dynamic linkage between different

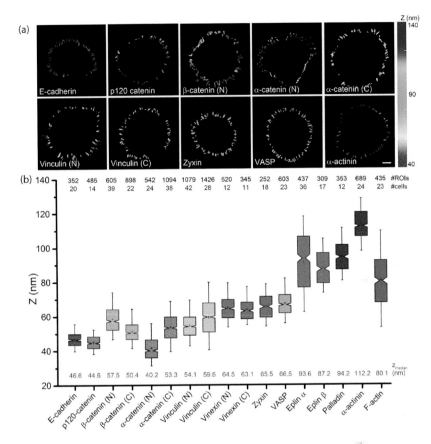

FIGURE 4.4 Protein stratifications in MDCK cadherin-based adhesions. (Figure 2 from[25] with minor modifications). (a) Topographic maps of protein z-positions (nanometers) (a) and notched box plots for the z-position of the indicated proteins (b) in E-cadherin-based adhesions of MDCK cells. The color bars in (a) indicate the z-position relative to the substrate surface. Scale bars, 10 μm. Notched box plots in (b) indicate first and third quartiles, median and confidence intervals; whiskers, 5th and 95th percentiles. Median zcenter values are indicated below each box plot (red). *n* values are shown above each box plot and indicate the numbers of adhesions (number of ROIs, black). Numbers of cells are indicated in blue. (Legend for the figure has been included as in the original paper[25] with minor modifications.)

shafts of a mechanical engine, differential friction caused by proteins' conformational changes engages the adhesive complex with actomyosin contractile system, effectively acting as a "molecular clutch" with a similar mechanism to FAs.[44]

The ease access obtainable by SAIM made possible to analyze the nanoscale ultrastructure of cadherin-mediated adhesion and their dynamics, to better appreciate the function–structure relationship, and to understand how cell-cell contacts are formed, maintained, regulated, and reinforced to perform vital biological functions.

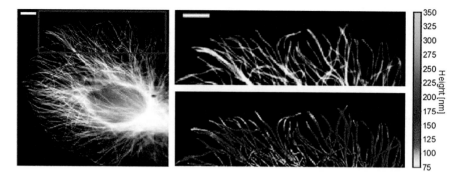

FIGURE 4.5   Scanning angle interference imaging of microtubules (Figure 2a from[22] with no modifications). Epifluorescence image (left) of fluorescently labeled microtubules (with tubulin-specific primary antibody and Alexa488-conjugated secondary antibody) in a fixed epithelial cell. The red box marks the region of interest (ROI) shown in the right panels; epifluorescence (top) and the corresponding scanning angle interference image, showing height reconstruction (bottom), are shown. Scale bars, 5 μm. Height is reported as the absolute distance of the fluorescent structure above the silicon oxide surface. (Legend for the figure has been included as in the original paper[22] with minor modifications.)

4.7.3   Axial Feature of Microtubules

In the same paper from Weaver's group[22] as in Section 4.7.1, the authors imaged various cellular structures, including but not limited to fluorescently labeled microtubules (Figure 4.5). Scanning angle interference images of fluorescently labeled microtubules in fixed epithelial cells showed a downward bending of the microtubule network in the lamella, with heights of ~ 70–350 nm. This result is extraordinary because this range of height goes beyond the working range offered by other high axial precision techniques using TIRF (Figure 4.5).

4.7.4   Axial Feature of Glycocalyx

Glycocalyx is a sugary film composed of multifunctional glycans and glycoproteins that coat every living cell in our body. The upregulated expression of glycoproteins plays a major role in the development of aggressive, lethal cancers.[45] In two recent publications,[46,47] SAIM has been used to map changes in the glycocalyx thickness induced by engineering its chemical and physical properties to understand how these sugars spatially configure the machinery of signal transduction. The results of such studies show that a bulky glycocalyx (due to overexpression of bulky glycoproteins) facilitates integrin clustering and by the application of tension, induces alteration in integrins' state. This explains how a bulky glycocalyx induces metastasis by mechanically modulating the function of cell-surface receptors in tumor cells and it is key to understanding how the spatial configuration of these sugars modifies signaling in the context of cancer.

4.8   CONCLUDING REMARKS

In conclusion, scanning angle interference microscopy is a valuable resource for biologists who are looking for a microscopy methodology that can enable the observation of nanoscale topographical features of complex supramolecular cellular structures. It provides unprecedented detail in the axial direction. It can be used for live cell imaging and thus it is suitable for observing time-dependent dynamics. It is applicable to nanoscale imaging of subcellular components on mechanically tunable substrates. All these characteristics, plus the possibility to readily implement SAIM on commercially available TIRF microscopes, distinguish SAIM from popular superresolution imaging techniques. Its applicability to studying biological structures and dynamics with nanoscale precision for complexes proximal to the cell membrane has already been well demonstrated. Now, the venue is open for its potential applications in the biomedical field.

ACKNOWLEDGMENTS

C.B. and A.R. are supported by Concurso de Investigación Interdisciplinaria 2018 awarded to C.B and A.R. (Semilla n. ii180002). T.R. is supported by Fondecyt Iniciación (n. 11161046). C.B., A.R and T.R. are supported by CONICYT/PIA/Anillo de Investigación en Ciencia y Tecnología n. ACT192015.

REFERENCES

1. National Research Council, Committee on Research Opportunities in Biology. *Opportunities in Biology* (Washington, DC: The National Academies Press, 1989).
2. Abbe, E. Note on the proper definition of the amplifying power of a lens or a lens-system. *J Royal Microsc Soc* **4**, 348–351 (1884).
3. Abbe, E. Beitrage zur Theorie des Mikroskops und der mikroskopischen Wahrnehmung. *Schultzes Arc f Mikr Anat* **9**, 413–468 (1873).
4. Rayleigh, J. On the theory of optical images, with special reference to the microscope. *London, Edinburgh, Dublin Philos Mag J Sci* **42**, 167–195 (1896).
5. Huang, B., Bates, M. & Zhuang, X. Super-resolution fluorescence microscopy. *Annu Rev Biochem* **78**, 993–1016 (2009).
6. Rust, M.J., Bates, M. & Zhuang, X. Sub-diffraction-limit imaging by stochastic optical reconstruction microscopy (STORM). *Nat Methods* **3**, 793–795 (2006).
7. Betzig, E. et al. Imaging intracellular fluorescent proteins at nanometer resolution. *Science* **313**, 1642–1645 (2006).
8. Hell, S.W. & Wichmann, J. Breaking the diffraction resolution limit by stimulated emission: Stimulated-emission-depletion fluorescence microscopy. *Opt Lett* **19**, 780–782 (1994).
9. Bertocchi, C., Goh, W.I., Zhang, Z. & Kanchanawong, P. Nanoscale imaging by superresolution fluorescence microscopy and its emerging applications in biomedical research. *Crit Rev Biomed Eng* **41**, 281–308 (2013).
10. Sahl, S.J., Hell, S.W. & Jakobs, S. Fluorescence nanoscopy in cell biology. *Nat Rev Mol Cell Biol* **18**, 685–701 (2017).

11. Galbraith, C.G. & Galbraith, J.A. Super-resolution microscopy at a glance. *J Cell Sci* **124**, 1607–1611 (2011).
12. Mockl, L., Lamb, D.C. & Brauchle, C. Super-resolved fluorescence microscopy: Nobel Prize in chemistry 2014 for Eric Betzig, Stefan Hell, and William E. Moerner. *Angew Chem Int Ed Engl* **53**, 13972–13977 (2014).
13. Deschout, H. et al. Precisely and accurately localizing single emitters in fluorescence microscopy. *Nat Methods* **11**, 253–266 (2014).
14. Shtengel, G. et al. Interferometric fluorescent super-resolution microscopy resolves 3D cellular ultrastructure. *Proc Natl Acad Sci U S A* **106**, 3125–3130 (2009).
15. Mondal, P.P. Temporal resolution in fluorescence imaging. *Front Mol Biosci* **1**, 11 (2014).
16. Gustafsson, M.G. Surpassing the lateral resolution limit by a factor of two using structured illumination microscopy. *J Microsc* **198**, 82–87 (2000).
17. Heintzmann, R.G., MGL. Subdiffraction resolution in continuous samples. *Nat Photonics* **3**, 362–364 (2009).
18. Lambacher, A. & Fromherz, P. Fluorescence interference-contrast microscopy on oxidized silicon using a monomolecular dye layer. *Appl Phys A-Materials Sci Proc* **63**, 207–216 (1996).
19. Lambacher, A.F., P. Luminescence of dye molecules on oxidized silicon and fluorescence interference contrast microscopy of biomembranes. *J Opt Soc Am B: Opt Phys* **19**, 1435–1456 (2002).
20. Braun, D. & Fromherz, P. Fluorescence interference-contrast microscopy of cell adhesion on oxidized silicon. *Appl Phys A-Materials Sci Proc* **65**, 341–348 (1997).
21. Ajo-Franklin, C.M., Ganesan, P. V. & Boxer, S. G. Variable incidence angle fluorescence interference contrast microscopy for z-imaging single objects. *Biophys J* **89**, 2759–2769 (2005).
22. Paszek, M.J. et al. Scanning angle interference microscopy reveals cell dynamics at the nanoscale. *Nat Methods* **9**, 825–827 (2012).
23. Bailey, B., Farkas, D.L., Taylor, D.L. & Lanni, F. Enhancement of axial resolution in fluorescence microscopy by standing-wave excitation. *Nature* **366**, 44–48 (1993).
24. Liu, J. et al. Talin determines the nanoscale architecture of focal adhesions. *Proc Natl Acad Sci USA* **112**, E4864–E4873 (2015).
25. Bertocchi, C. et al. Nanoscale architecture of cadherin-based cell adhesions. *Nat Cell Biol* **19**, 28–37 (2017).
26. Ou, G.Q. et al. Visualizing mechanical modulation of nanoscale organization of cell-matrix adhesions. *Integrative Biol* **8**, 795–804 (2016).
27. Kiessling, V. & Tamm, L.K. Measuring distances in supported bilayers by fluorescence interference-contrast microscopy: Polymer supports and SNARE proteins. *Biophys J* **84**, 408–418 (2003).
28. Braun, D. & Fromherz, P. Fluorescence interferometry of neuronal cell adhesion on microstructured silicon. *Phys Rev Lett* **81**, 5241–5244 (1998).
29. Stock, K. et al. Variable-angle total internal reflection fluorescence microscopy (VA-TIRFM): Realization and application of a compact illumination device. *J Microsc* **211**, 19–29 (2003).
30. Axelrod, D. Cell-substrate contacts illuminated by total internal reflection fluorescence. *J Cell Biol* **89**, 141–145 (1981).
31. Carbone, C.B., Vale, R.D. & Stuurman, N. An acquisition and analysis pipeline for scanning angle interference microscopy. *Nat Methods* **13**, 897–898 (2016).
32. Toworfe, G.K. et al. Effect of functional end groups of silane self-assembled monolayer surfaces on apatite formation, fibronectin adsorption and osteoblast cell function. *J Tissue Eng Regen Med* **3**, 26–36 (2009).

33. Day, R.N. & Davidson, M.W. The fluorescent protein palette: Tools for cellular imaging. *Chem Soc Rev* **38**, 2887–2921 (2009).
34. Hamers-Casterman, C. et al. Naturally occurring antibodies devoid of light chains. *Nature* **363**, 446–448 (1993).
35. DuFort, C. & Paszek, M. Nanoscale cellular imaging with scanning angle interference microscopy. *Methods Cell Biol* **123**, 235–252 (2014).
36. Nyquist, H. Certain factors affecting telegraph speed. *Bell Sys Techn J* **3**, 324–346 (1924).
37. Iskratsch, T., Wolfenson, H. & Sheetz, M.P. Appreciating force and shape-the rise of mechanotransduction in cell biology. *Nat Rev Mol Cell Biol* **15**, 825–833 (2014).
38. Lim, C.T., Bershadsky, A. & Sheetz, M.P. Mechanobiology. *J R Soc Interface* **7**(Suppl 3), S291–S293 (2010).
39. Engler, A.J., Sen, S., Sweeney, H.L. & Discher, D.E. Matrix elasticity directs stem cell lineage specification. *Cell* **126**, 677–689 (2006).
40. Ou, G. et al. Visualizing mechanical modulation of nanoscale organization of cell-matrix adhesions. *Integr Biol (Camb)* **8**, 795–804 (2016).
41. Prabhune, M., Rehfeldt, F. & Schmidt, C.F. Molecular force sensors to measure stress in cells. *J Phys D-Appl Phys* **50**, 233001 (2017).
42. Kanchanawong, P. et al. Nanoscale architecture of integrin-based cell adhesions. *Nature* **468**, 580–584 (2010).
43. Zaidel-Bar, R. Cadherin adhesome at a glance. *J Cell Sci* **126**, 373–378 (2013).
44. Case, L.B. & Waterman, C.M. Integration of actin dynamics and cell adhesion by a three-dimensional, mechanosensitive molecular clutch. *Nat Cell Biol* **17**, 955–963 (2015).
45. Kang, H. et al. Cancer cell glycocalyx and its significance in cancer progression. *Int J Mol Sci* **19**, 2484 (2018).
46. Paszek, M.J. et al. The cancer glycocalyx mechanically primes integrin-mediated growth and survival. *Nature* **511**, 319–325 (2014).
47. Shurer, C.R. et al. Genetically encoded toolbox for glycocalyx engineering: Tunable control of cell adhesion, survival, and cancer cell behaviors. *ACS Biomater Sci Eng* **4**, 388–399 (2018).

5 Atomic Force Microscopy of Biomembranes

A Tool for Studying the Dynamic Behavior of Membrane Proteins

Yi Ruan, Lorena Redondo-Morata, and Simon Scheuring

CONTENTS

5.1   INTRODUCTION

This chapter starts with an overview of the atomic force microscopy (AFM) technique, gives detailed explanations of the operational fundamentals and the recent developments of high-speed atomic force microscopy (HS-AFM), describes the use of HS-AFM for biological membrane proteins research and introduces typical results obtained by HS-AFM in biology, with a particular focus on membrane-embedded and membrane-associated proteins.

Atomic force microscopy (AFM) has evolved into an established powerful tool for studying biological samples under nearly physiological conditions ranging from cells to single molecules. AFM imaging generates topographical maps of biological surfaces with a spatial resolution of about 1 nm, and has therefore become a complementary technique to X-ray crystallography, NMR and electron microscopy to elucidate biological structures. In contrast to the other high-resolution techniques, AFM provides a way to study molecules under near physiological conditions, e.g., in liquid environment, under ambient pressure and temperature, and without the requirement of any labeling or staining. Hence, it allows the direct observation of structure–function relationships on intact, native samples. However, conventional AFM with image acquisition rates of several minutes impeded the analysis of relevant biological dynamic processes. During the last decade, efforts have been made to increase the AFM scan rate. Thanks to the HS-AFM development, it is now possible to acquire movies at rates of about 10 frames per second and about 200 data points squared per frame.

Membrane proteins and lipids are the main component of biological membranes. The characterization of the supramolecular organization and dynamics of membrane proteins is crucial for the understanding of the structure and function of biomembranes. AFM is now a unique tool for the investigation of membrane proteins at single molecule level resolution in a native-like environment. Combining all the advantages of AFM with a high temporal resolution in the millisecond range. HS-AFM allows both real-space and real-time visualization of the protein dynamics, providing an excellent platform to establish structure–function relationships and to determine biophysical parameters of single molecules previously inaccessible by other techniques.

5.2   OVERVIEW OF ATOMIC FORCE MICROSCOPY (AFM)

Atomic force microscopy (AFM) was developed in 1986, allowing users to resolve single atoms on non-conductive solid surfaces.[1,2] After the development of the fluid cell,[3] AFM became a powerful tool in structural biology and biophysics.[4] For its operation, oscillatory modes such as "tapping mode" AFM are the most frequently used when operated in ambient conditions or in liquids for biological research because the cantilever tip-sample interaction is minimized in the normal direction, these modes also exerts little lateral force or friction on soft fragile samples, compared with other modes such as contact mode, where the tip remains in permanent contact with the sample. The operational principle of the AFM technique consists of scanning the sample surface with a very sharp tip which is attached to the end of a flexible cantilever that serves as a force sensor. The position of the cantilever is detected by an Optical Beam Detection (OBD) system. The cantilever is driven to oscillate up and down at or near its resonance frequency, a constant amplitude of the cantilever oscillation is kept in order to maintain a constant force applied to the sample using a feedback controller. The resolution that can be achieved is limited by the compliance of the sample and the geometry of the tip. By controlling the aqueous conditions such as pH and ionic strength and applying minimal forces between the tip and the surface, AFM is able to resolve single molecules in a close-to-native state with typical lateral resolutions of ~1 nm and vertical resolution of ~0.1 nm, even sub-nanometer resolution have been reported.[5-7] However, conventional AFM image acquisition rates are in the minute's time range, and thus provide only static snapshots of dynamic biological processes.

5.2.1 Principle of High-Speed Atomic Force Microscopy (HS-AFM)

With the progresses in structural biology, an increasing demand for studying the dynamics of biological samples prompted to steady improvements of the AFM technology, resulting in a noticeably increased image acquisition rate. Around the year 2000, the first reports were made about AFMs using short cantilevers with increased scan rate.[8-10] Further key adaptations for increasing the scan speed mainly concerned the miniaturization of the moving parts of the AFM setup, because the speed at which these elements can be moved with accuracy is proportional to the inverse of the square root of their mass, improved optical detection and electronics.

First, light microfabricated cantilevers of short lengths, typically between 5 µm and 10 µm, were developed. They have typical resonance frequencies between 500 kHz and 1 MHz in liquid,[9] keeping a low spring constant of ~200 pN/nm, and are suitable for biological specimens. Another advantage of such short levers is the sensitivity of the OBD system,[11] consisting of a laser focused on and reflected by the backside of the cantilever and detected by a split photodiode: The OBD detects the angular deflection and thus the same absolute deflection results in a larger angular deflection signal with a short cantilever. Second, scanners composed of individual monolithic X, Y, and Z piezoelectric elements have been designed. These scanners fulfill several criteria, such as high resonance frequency, few resonant peaks at similar frequencies, small quality factor, small cross-talk, etc., while a reasonable scan range sufficient for bio-molecular surface analysis is assured.[12,13] Crucial for fast surface contouring is a small Z-piezo with a high resonance frequency of ~150 kHz,[13,14] which is driven by a dynamic feedback-controller that alters the gain parameters to follow steep surface features.[15]

With the above-described key elements and some more adaptations, HS-AFM is now able to acquire movies at ~10 frames per second with ~200 data points squared per frame.[8] For bio-molecular observations where typically a scanned area of at least 100 nm is required, the achievable resolution is ~1 nm, allowing the visualization of single molecule processes with sub-molecular resolution. Overall, these developments allowed, for the first time, the observation of single molecules at work under close to physiological conditions, providing an advanced way for researchers to establish structure–function relationships and to determine biophysical parameters of single molecules previously inaccessible by other techniques.

5.2.2 INSTRUMENTATION OF HS-AFM

The first theoretical analysis of the scan speed limits of AFM was reported by Butt et al. 1993, focusing on the relationship between the cantilever's mechanical properties and the imaging speed.[16] In the following years, remarkable efforts were invested to increase the AFM imaging speed in the groups of Hansma,[10,11,17,18] Quate,[19,20] and Ando.[21,2] In 2001, Ando et al. reported a complete high-speed AFM system, consisting of a high-speed scanner, small cantilevers (resonant frequency 450–650 kHz in water; spring constant, 150–280 pN/nm), fast electronics (PID feedback), and an Optical Beam Deflection (OBD) detector adapted to the small cantilevers. A schematic depiction of tapping-mode AFM[21–23] is shown in Figure 5.1.

5.2.2.1 Sample Scanner

The sample scanner must fulfill several critical requirements: (1) rapid response to the impulsive force, (2) high resonant frequencies, (3) a small number of resonant peaks in a narrow range of frequencies, (4) sufficient maximum displacements, (5) small interference (crosstalk) between the three axes (X, Y and Z) and (6) low quality factors.[22] A symmetrical and compact sample scanner design has been shown to be well adapted for biological HS-AFM, as shown in Figure 5.2.

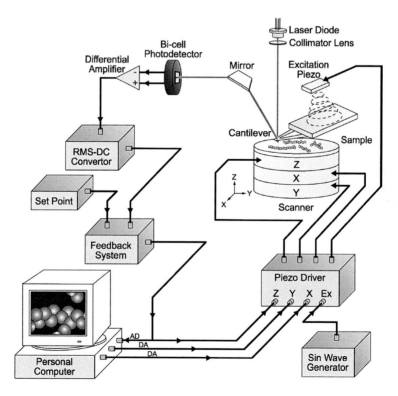

FIGURE 5.1 Schematic presentation of a tapping-mode AFM system. (Adapted from Ando, T. et al., *e-J. Surf. Sci. Nanotechnol.*, 3, 384–392, 2005. © 2016 Ando et al. CC BY 4.0.)

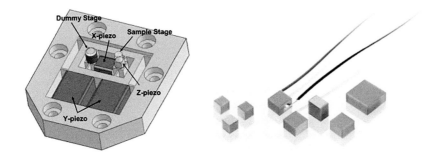

FIGURE 5.2 Left: Sketch of a high-speed HS-AFM scanner amenable for movie acquisition of biological molecules. (Adapted from Ando, T. et al., *Progr. Surf. Sci.*, 83, 337–437, 2008. With permission.) Right: Z-actuators used in HS-AFM. (Adapted from the product website of Physik Instrumente company. https://www.physikinstrumente.com/. With permission.)

Due to its large mechanical damping coefficients and large ratios of Young's modulus to density, magnesium is chosen to be the material for these types of scanners. To move the sample in three dimensions, three piezoelectric actuators are aligned perpendicularly to each other. This type of piezo is commercially available from Physik Instrumente (PI) GmbH & Co. KG (Germany). The dimensions (W × L × H) of Z-actuator are $2 \pm 0.1 \times 3 \pm 0.1 \times 3 \pm 0.1$ mm^3, which has a resonant frequency of >600 kHz in free oscillation and with a maximum displacement of 2.2 μm (±20%). The gaps in the scanner are filled with a soft polyurethane elastomer to passively suppress the low-frequency vibrations introduced by the quick displacements of the actuators.

5.2.2.2 Small Cantilever

Ultra-short cantilevers for HS-AFM are now commercially produced by the companies Nanoworld and Olympus. Cantilevers are usually made of a quartz-like material, and typical dimensions are (L × W) are $7 \pm 1 \times 2 \pm 0.5$, and 0.08 ± 0.02 μm of thickness. A ~20 nm gold layer is deposited on both sides of the cantilever in order to enhance the reflectance of the laser beam. The sharp and wear resistant tips are made through electron beam deposition in a SEM and plasma-sharpening techniques. Overall, this short cantilever shown in Figure 5.3 has a nominal spring constant of ~0.15 N/m and a resonance frequency of ~0.6 MHz in liquid. Besides the advantage of allowing high imaging rate, small cantilevers have other benefits, including low noise density,[24] and high OBD detection sensitivity.[22] The tip has an apex radius of several nanometers, which can be sharpened through plasma etching in argon or oxygen gas down to ~1 nm. This is also the common method used in the course of HS-AFM experiments for tip reuse and/or cleansing from contaminations.

5.2.2.3 Fast Electronics (PID Feedback)

To achieve fast imaging speed, the Proportional-Integral-Derivative (PID) feedback should respond quickly enough to keep the tip force constant during scanning. In amplitude modulation mode HS-AFM, the cantilever tip intermittently taps the sample as the tip scans over the sample surface. This mode is suitable for imaging

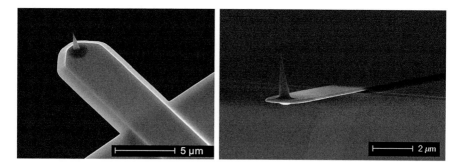

FIGURE 5.3 Image of an ultra-short cantilever (USC-F1.2-k0.15) used for HS-AFM. Left: cantilever 3D view close-up. Right: sharp tip side view. (Adapted from the product website of Nanoworld company, https://www.nanoworld.com/Ultra-Short-Cantilevers-USC-F1.2-k0.15. html. With permission.)

fragile samples such as biological macromolecules, because the vertical oscillation of the cantilever reduces lateral forces between the tip and the sample. In order to observe the weak biomolecular interactions at high resolution, the tapping force (vertical force) must be minimized. Various efforts have been carried out. One way is to reduce the spring constant of small cantilevers; however, it is difficult to make it compatible with high resonant frequency for fast imaging. Another way is to set a large set-point, which means that the amplitude set point A_s (the peak-to-peak oscillation amplitude to be maintained during scanning) should be set very close to the free oscillation peak-to-peak amplitude $2A_0$. If A_s is close to $2A_0$ prolongation the feedback runs into the risk of losing the contact with the sample, so-called "parachuting," because the PID circuit produces a small feedback signal that is proportional to the saturated error signal when the tip is out of contact and oscillated at A_0. The feedback cannot be increased to shorten parachuting time because a larger feedback induces overshoot at uphill regions of the sample, resulting in the instability of the feedback operation.[15,22] To overcome this difficulty, a dynamic feedback controller has been developed, which automatically alters the feedback gain parameters depending on the cantilever oscillation amplitude during scanning.[15] The principle of the dynamic PID control is depicted in Figure 5.4.

Using the dynamic PID, a threshold level A_{upper} is set between A_s and $2A_0$. When the cantilever's peak-to-peak oscillation amplitude A_{p-p} exceeds A_{upper}, the differential signal $(A_{p-p}-A_{upper})$ is amplified and added to the error signal. The error signal that contains an extra signal is fed to the conventional PID which in turn results in drastically shortened parachuting periods, thus achieving a much quicker feedback response. The similar operation could be made in the case of a threshold of A_{lower}.

5.2.3 DEVELOPMENT OF HS-AFM

5.2.3.1 Buffer Exchange Pumping System

Biological processes such as gating of ion channels, translocation mechanism of transporters, ligand-recognition and signaling events of receptors, and so on are

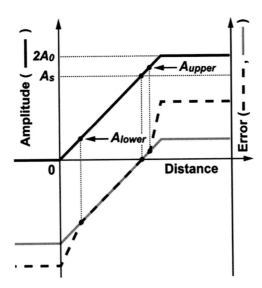

FIGURE 5.4 Schematic diagram showing the principle of the dynamic PID control. Solid line: an amplitude-distance curve; gray line: an error signal used in the conventional PID control; and broken line: an error signal used in the dynamic PID control. (Adapted from Kodera N. et al., *Rev. Sci. Instr.*, 77, 2006. With permission.)

triggered by different stimuli such as voltage, temperature, ions, pH or by recognition and binding of small molecules. Accordingly, the addition or removal of ions or ligands during the experimentation is crucial to mimic physiological processes and to study the workings of the molecule of interest. Typically, in all AFM experiments (conventional and HS-AFM) the measuring condition is changed by exchanging the buffer by thorough washing of the sample or by adding the trigger molecules into the measuring bath by pipetting high concentrated solutions into particular inlets. Even though this is a straightforward protocol, it has some disadvantages and can cause some considerable problems like: (1) probable loss of the identical image position, and the corresponding loss in direct comparison before and after the stimulation; (2) image perturbations; (3) probable damage of the sample due to shifts in the imaging amplitude caused by perturbations; (4) tip contamination due to the abrupt buffer addition; (5) no gradual, continuous exchange is possible; (6) impossibility of applying concentration gradients; (7) experiment is often feasible only in one direction (e.g., the ligand being added, but not removed when complete removal of ligands would require the disassembly and rinsing of the sample chamber and thus a position loss); and (8) difficulties in capturing very sensible and fast reactions.

To overcome these problems and to enable a slow, gradual controlled change in measuring conditions, a bio-enhanced HS-AFM was developed through the integration of a high precision pumping system.[25–27] A constant-pressure and constant flow pump is connected via tubes using particular in- and outlets to the HS-AFM fluid chamber forming a closed circuit. By simultaneously injecting the buffer solution on one side and removing it on the other side of the fluid cell, complex buffer composition exchanges are possible with rates ranging from 0.01 to 1 μL/s. This implemented and optimized buffer exchanging system is used to gradually change the ligand concentration from one condition to the other (high to low ligand concentration or vice versa) while simultaneously recording high-resolution movies without perturbations. However, to test the reversibility of a biological process, and to exclude probable unspecific effects, it is often desirable to switch back and forth between different conditions multiple times. Moreover, in case of multiple binding sites for a second different ligand, or performing a specificity proof using another ion, inhibitors, low-affinity ligands or simply testing any different third condition it can be useful to have another buffer exchanging path. Consequently, to achieve repeated buffer exchange or to add a third buffer condition, an additional syringe is applied to the buffer exchange pumping system. The tube runs in parallel with the first injection syringe, and is introduced into the pathway of the first injection tube using a T-type switch as shown in Figure 5.5.

In this high precision buffer exchange system, the "buffer in" and "buffer out" channels are placed in proximity to the HS-AFM cantilever (central element inside the fluid cell). The fluid cell contains initially ~150 μL of buffer 1 (blue) that is connected to the receiving syringe. Buffer 2 (red) is the first buffer to be injected into the fluid cell. Note, the slight gap between buffer 2 and the fluid cell, representing a tiny bubble in the tubing to avoid involuntary mixing of buffer 2 before activation of the pumping system. By the use of a switch in the tubing system, buffer 3 (green) can later be injected into the fluid cell. Buffer 3 can be identical to buffer 1, allowing for a reversed experiment.[28] The integration of such a buffer exchange system to the HS-AFM allows for structural titration experiments system.[25–27] For fast environmental changes photo-activated uncaging of caged compounds.[25–27]

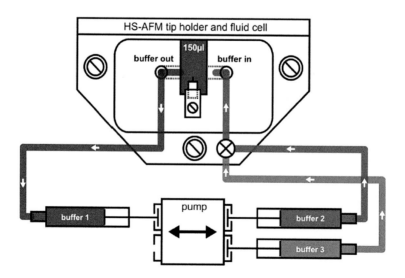

FIGURE 5.5 Schematic representation of the coupling of the buffer exchange system to the HS-AFM fluid cell.

5.2.3.2   Temperature Control System

Most of AFM experiments are carried out under room temperature. However, most bio-logical process are temperature-dependent. For example, dynamics of enzymatic reactions, protein unfolding, membrane protein diffusion, lipid bilayer phase transition,[29,30] sensitivity of certain ion channels,[31] etc., are temperature-dependent. Thus, a temperature control device was developed and integrated with HS-AFM in order to achieve real-time observations of temperature-sensitive processes, in Figure 5.6. The bottom glass plate of a HS-AFM cantilever holder is coated with a tungsten deposit using a high vacuum sputter coater in argon environment. The tungsten coat has a typical thickness of 35 nm and a resistance value of 200 Ω. Both ends of the glass plate are connected to a DC power supply via cables connected with conductive glue (Figure 5.6a). The tungsten-coated glass plate is then assembled into the cantilever holder of the HS-AFM. For the measurement of temperature during imaging, a small thermocouple of 0.075 mm in diameter is placed underneath and in contact with the cantilever chip, 1 mm away from the HS-AFM tip, connected to a thermometer (Figure 5.6b). Contact of the thermocouple to the cantilever chip and its proximity to the tip assure rapid and precise assessment of the actual temperature at the point of investigation. The temperature is recorded every second, in accordance with the frame rate (Figure 5.6c).

The development of this temperature-controlled HS-AFM completes a series of developments that allows the environmental conditions to be controlled during HS-AFM operation, which makes the HS-AFM a versatile tool for thermometric analysis of dynamic structural biochemistry.

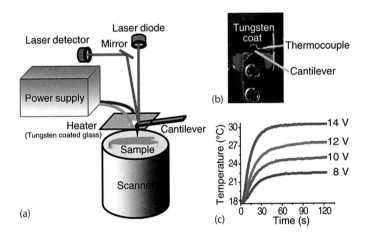

FIGURE 5.6   Development of temperature-controlled HS-AFM. (a) Schematic illustration of the heating system for HS-AFM represented by a Tungsten coated glass used as resistor. (b) Assembly of the tungsten-coated glass, the cantilever, and the thermocouple. The thermocouple is attached underneath the cantilever chip close to the HS-AFM tip to measure local temperature at the imaging area. (c) Temperature as function of input voltage to the tungsten-coated glass. The temperature inside the imaging chamber filled with 100 μL Milli-Q water is measured by the thermocouple underneath the cantilever. The temperatures measured every 1 seconds at different values of input voltage (here from 8 to 14 V) are plotted against time. (Adapted from Takahashi, H. et al., *Small*, 16, 6106–6113, 2016. With permission.)

5.2.3.3   Integrating Light-Microscopy with HS-AFM

Light microscopy, including fluorescent microcopy, is the most frequently used tool to characterize the supramolecular organization of cellular membranes. Although light microscopy presents cell structural analysis at a resolution inferior to HS-AFM, the images of optical microscopy such as fluorescent images can reveal the cell interior, not only the membrane surface of the cells. Thus, HS-AFM and optical microscopy are combined in order to take advantage of light microscopy's large-scale overview imaging capacities and its power to analyze fluorescence signals from the labeled proteins of interest.[32] The scheme for this method is shown in Figure 5.7.

In details, to integrate the M into the light microscopy path into HS-AFM, a super luminescent diode (SLD) with emission wavelength at 750 nm (labeled 1 in Figure 5.7) is used. Two light sources, one for fluorescence microscopy (labeled 2

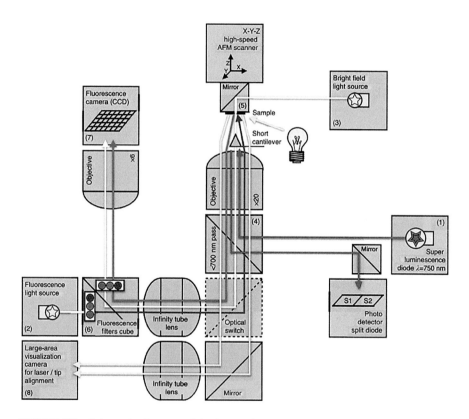

FIGURE 5.7   Schematic diagram of the integration of light microscopy into the HS-AFM setup. The essential elements integrated are: a 750-nm SLD for cantilever position detection (1), two light microscopy sources (2 and 3), a 700-nm low-pass beam splitter (4), a miniaturized mirror sample support (5), a fluorescence filter cube beam splitter (6) and a high-resolution CCD camera (7). A large-area visualization camera allows the alignment of the laser on the HS-AFM tip (8). Several elements (blue) are physically uncoupled from the HS-AFM body (gray). (Adapted from Colom et al., *Nat. Commun.*, 4, 2155, 2013. © 2013 Scheuring et al. CC BY 4.0)

in Figure 5.7) and one for bright field microscopy (labeled 3 in Figure 5.7) and pathways are added into original HS-AFM setup. A short pass-wavelength dichroic filter (pass: <700 nm, labeled 4 in Figure 5.7) and an optical switch 45° mirror driven by a linear motor stage (DDSM100, ThorLabs) and controlled by a servo (TBD001, ThorLabs) are integrated for switching between various optical paths. The lateral light injection sample holder (labeled 5 in Figure 5.7) is composed of two prisms with a 100-nm aluminum-coated interface glued together by cyanoacrylate. The fluorescence filter cube (labeled 6 in Figure 5.7) allows filtering of specific wavelengths for the excitation and light recovery of fluorescent dyes. Finally, an infinity tube lens unit is integrated, and a X6 macro zoom lens focuses the light microscopy signals into a CCD camera (labeled 7 in Figure 5.7).

This HS-AFM-light microscopy hybrid setup allows the biologist to observe the structure-function dynamics of membrane proteins in real-time following fluorescence-based targeting of zones of interest on cells, and therefore pave novel avenues for cell biology applications of HS-AFM.

5.2.4   Use of HS-AFM: Limitations and Advantages

By focusing on the principle and the research activities on HS-AFM in the past decade, we briefly introduced the instrumentation and the remarkable developments of HS-AFM applied to the study of biological processes.

Overall, the advantage of using HS-AFM is that we can directly observe the structural dynamics and dynamic processes of biomolecules at subsecond temporal and submolecular spatial resolutions without disturbing their function. The information provided by HS-AFM is inaccessible to other approaches. The combination of HS-AFM with high-resolution techniques, e.g., X-ray crystallography, NMR and electron microscopy could offer great insights into how the molecules function, and interpret straightforwardly the dynamics of molecules without intricate analyses.

However, the current HS-AFM has several limitations. (1) To achieve fast scanning speed with high resolutions, the scan field is limited to micrometers and even tens of nanometers for very fast scanning. (2) The forces applied by the tip contacting the sample must be controlled with great precision, and while imaging molecular systems on supports can be performed without inducing molecular damages, it is currently impossible to observe single molecules on the surface of soft life cells. (3) The buffer-exchange rate is limited to 15 µL/minute due to solution redistribution introduced by buffer injection/extraction. (4) The temperature variation range is limited to around 20 degrees C because of mechanical drift and evaporation introduced by the buffer temperature change. The ongoing efforts to overcome these limitations could be found in references,[33–44] but we will not further describe them in this chapter.

5.3   HS-AFM OF BIOLOGICAL MEMBRANES

The main objective of HS-AFM experiments of biological membranes is to characterize the topographical structure and the dynamics of membrane proteins under liquid environment at high temporal and spatial resolutions. Using HS-AFM and the developments within the technique, it is possible to answer relevant biological questions about the structure–function relationships on model and native biomembranes, without labeling and with a spatial resolution of ~1 nm, far beyond the optical diffraction limit. However, as described above, it is possible, although difficult, to scan the tip on the soft surface, such as cellular membranes, because of the surface deformation introduced by the tip-sample interaction. Up to date, most HS-AFM experiments of biological membranes are carried out on a mica surface using artificial model membranes. To investigate molecular mechanisms associated with the membrane components, Supported Lipid Bilayers (SLBs) are a good choice. In this section, we will describe in detail the protocols to fabricate artificial SLBs on small mica disks using the vesicle fusion method. Furthermore, successful efforts have been made to reconstitute integral proteins in lipid bilayers and adopt them to HS-AFM experiments.[45]

5.3.1 COMPOSITION OF BIOLOGICAL MEMBRANES

Compartmentalizing chemical elements and simple compounds by membranes was probably a very early and important step in the development of life. A biological membrane is an enclosing or separating film that acts as a selectively permeable barrier within living cells.[46] In a cell membrane, it contains a variety of biological molecules, notably lipids and proteins. Lipid provides a fluid matrix for proteins to rotate and laterally diffuse, while proteins with their specific functions are responsible for various biological activities, from transmembrane transport to cell signaling, cell adhesion, etc. Proteins can be roughly divided into three main classes: globular proteins, fibrous proteins, and membrane proteins.[47] In this chapter, we focus on membrane proteins, which serve as receptors or provide channels for polar or charged molecules to pass through the hydrophobic core of the cell membrane. By using HS-AFM with its latest developments, the dynamics of proteins involved in such processes can be directly visualized and analyzed.

5.3.2 MEMBRANE PROTEINS

Membrane proteins are proteins that interact with or are inserted in biological membranes. Those embedded in cell membranes serve a critical purpose in the maintenance of many cellular functions, including signal transduction, cell integrity, intracellular and extracellular transport, cell-to-cell communication, etc. In principle, membrane proteins are mainly divided into two categories: peripheral membrane proteins, integral membrane proteins. The former may temporarily or permanently go through lipids, covalent link to lipids, or bind to other proteins in membranes, while the latter are permanently embedded in the membrane. The new features of HS-AFM enable us to analyze dynamics of membrane proteins, such as how single membrane proteins move, how they interact with other membrane proteins, their oligomeric state and conformational changes, etc., we will describe some of these studies in the following.

5.3.3 HS-AFM Experiments

5.3.3.1 HS-AFM Experiments of Peripheral Membrane Proteins

5.3.3.1.1 HS-AFM Imaging of Polypeptide Toxins

Polypeptide toxins, the third category of membrane proteins, are usually described as peripheral membrane proteins, such as listeriolysin-O, lysenin, α-hemolysin, etc. They are water-soluble, but can aggregate and associate reversibly or irreversibly with a lipid bilayer and become membrane-associated and membrane-integrated.[48] HS-AFM movie acquisition of a polypeptide toxin assembly on model membranes at sub-second temporal resolution allows the analysis of its membrane-perforating action in a series of well-defined experimental conditions, e.g., the dynamics of the protein when exposed to lipid bilayers containing varying amounts of cholesterol or to varying environmental pH, etc.

Step-by-step sample preparation protocol

- **Proteins:** The pure protein is aliquoted and is stored in a buffer at a suitable concentration.
- **Liposomes:** All lipids are dissolved in an organic solution of chloroform: methanol (3:1, vol:vol) to give a final concentration of several mM. An aliquot is poured into a glass vial and dried with clean nitrogen flow. The resulted lipid film is kept under reduced pressure overnight to remove trace amounts of organic solvent. Afterwards, the lipid film is hydrated with Milli-Q water to a final lipid concentration of hundreds of μM before five cycles of agitation of the vial for 1 minute each and then heating to ~70°C, which is well above the transition temperature of the lipid mixture. This procedure produces multilamellar vesicles, which are thereafter sonicated for 40 min in order to obtain Large Unilamellar Vesicles (LUVs). After preparation, LUV suspensions are stored at ~4°C and used within a maximal duration of 10 days. During the preparation, the samples are protected from light to avoid unspecific damage from oxidation.
- **Supported Lipid Bilayers (SLBs):** SLBs are prepared by fusing LUVs on a mica support.[49] To form SLBs, 2 μL of LUVs are deposited on the surface of a 1.5 mm² freshly cleaved mica. The mica sheet is glued with epoxy to a quartz sample stage. After incubation in a humid chamber for 30–40 minutes, the sample is gently rinsed with milli-Q water and kept wet all the time.

5.3.3.1.1.1 HS-AFM Experiments

HS-AFM movies are acquired using an ultra-short cantilever (Type: USC-F1.2-k0.15, Nanoworld) with a nominal spring constant of $k = 0.15$ N/m, and a resonance frequency of $f = 0.6$ MHz in solution. Both the cantilever and the rinsed mica surface with the attached bilayers are placed into a imaging chamber of a volume of 120 μL. HS-AFM is operated in amplitude modulation mode. HS-AFM measurements are performed at room temperature. Small free and set-point a oscillation amplitudes of about 1 and 0.9 nm, respectively, are used, to achieve minimum tip-sample interaction. HS-AFM could be used for monitoring the process of SLB formation on the mica surface. Soluble protein toxins are added to a final concentration of a few hundreds of nM after identification of the membrane patches on the mica surface. Dynamic protein assembly on a model membrane are observed and recorded directly by HS-AFM.

5.3.3.1.1.2 Typical Results   Given the advantage of high spatial and temporal resolution of HS-AFM, the mechanism of Listeriolysin-O (LLO) action on SLBs have been studies in real time. LLO is a soluble protein of 56 kDa in mass and belongs to a family of cholesterol-dependent cytolysin (CDCs) proteins. It plays a crucial role during infection by *Listeria monocytogenes*. Ruan et al.[50] discovered that LLO-SLB interaction is pH-dependent. LLO is able to form arc pores and can damage lipid membranes as a lineactant. Lineactants are molecules that tend to accumulate in 1D in the phase boundaries, that is, adsorbing at the contact line which separate two 2D phases. This process leads to large-scale membrane defects that can help bacteria to escape from phagocytic vacuoles. The dynamic imaging of HS-AFM revealed a detailed understanding of the molecular action of LLO as depicted in Figure 5.8.

    Another interesting result is on the process of protein insertion via anomalous diffusion analysis of Lysenin, a 33 kDa protein extracted from the coleomic fluid of the earthworm *Eisenia fetida*. In a highly organized, crowded and clustered mosaic membrane of lipids and proteins, it has been difficult to correlate protein diffusion

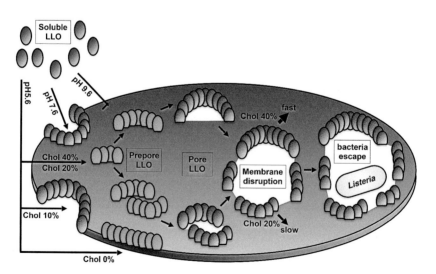

FIGURE 5.8   Schematic depiction of LLO membrane-disrupting action. LLO action depends on the cholesterol content in membrane and the environmental pH. LLO forms arc-shaped pores and creates large-scale defects as a lineactant for *Listeria* escape. Soluble LLO is represented in blue, pre-pore LLO in green, and pore LLO in orange, respectively. (Adapted from Ruan, Y. et al., *Plos Pathog.*, e1005597, 2016. © 2016 Ruan et al. CC BY 4.0)

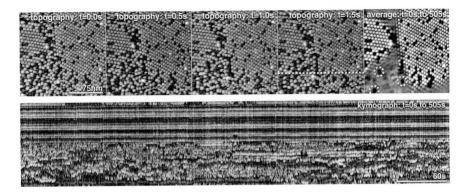

FIGURE 5.9 Lysenin dynamics is location-dependent. Upper: Four left panels: HS-AFM movie frames of lysenin in a sphingomyelin:cholesterol (1:1 weight ratio) bilayer. Right panel: Time-averaged frame displaying the positional stability and the high mobility of proteins in the solid and fluid domains (full false color scale: 10 nm). Lower: Kymograph (of the white dashed line in a). Stable and highly mobile molecules are in the top and the bottom while those switching between stability and high mobility (middle) are visible during the entire movie. (Reprinted with permission from Munguira, I. et al., *ACS Nano*, 10, 2584–2590, 2016. Copyright 2016 American Chemical Society.)

properties and domain formation and local molecular details because single molecule fluorescence microscopy does not allow observing directly unlabeled molecules in a crowded environment. Munguira et al.[51] presented a novel method to analyze lysenin in a highly crowded environment and documented coexistence of different diffusion regimes within one membrane using HS-AFM, in Figure 5.9.

5.3.3.1.2 HS-AFM Spectroscopy of Spectrin

Besides imaging, another major application of AFM is force spectroscopy (AFM-FS), a method for direct measurement of tip-sample interaction forces as a function of the displacement between the tip and the sample, thus force spectroscopy allowed to measure ligand-receptor binding and protein unfolding through recording of the forces experienced by the cantilever during tip approach-retract cycles. Spectrin, a long, thin, flexible rod-shaped protein that is ~100 nm in length, is a cytoskeletal protein and a peripheral membrane protein. It is the principal component of the cytoskeleton that underlies the cell membrane, the paradigm example of which is the red blood cell (RBC) membrane. It maintains the structural integrity and biconcave shape of the RBC membrane. It is located at the inner face of the plasma membrane, making connections between membrane anchors and the actin cortex meshwork and between actin filaments. Compared to conventional AFM-FS,[52,53] HS-AFM-FS with short cantilevers allows unfolding spectrin proteins at velocities up to 10^{-2} m/s.[54] This is faster than conventional AFM-FS by several orders of magnitude and reaches a rate comparable to steered molecular dynamics simulations (SMDS).[55,56] HS-AFM-FS combined with SMDS spectrin unfolding, revealed the protein's mechanical properties, multiple unfolding pathways, intermediate folding states, etc.

Step-by-step sample preparation protocol

- **Proteins:** The DNA plasmid with a cDNA coding for ybbR, four repeats of Spectrin and XMod-dockerin-III, is transformed into BL21 (DE3) cells. The BL21 cells are inoculated in 50 mL of LB medium containing 50 ng/mL kanamycin and incubated overnight at 37°C with constant shaking. Next day the culture is transferred into 1 L of fresh LB medium containing 100 ng/mL kanamycin and incubated at 37°C with shaking until an OD_{600} of 0.6. To induce protein expression, isopropyl β-D-1-thiogalactopyranoside (IPTG) is added to the culture medium. The cells are incubated at 20°C with shaking overnight. On day 3, the overexpressing cells are centrifuged at 5000 g for 10 minutes. The pellet is collected and is resuspended in 10 mL of a HEPES-buffered saline (20 mM HEPES-NaOH, pH 7.5, 100 mM NaCl). After addition of 1 mM phenylmethylsulfonyl fluoride (PMSF), the bacteria are sonicated using a probe sonicator on ice (3 times for 20 seconds with 30 second intervals). Triton X-100, DNase, and RNase are added into the suspension at 0.1%, 10 μg/mL, and 20 μg/mL final concentrations, respectively. The mixture is incubated for another 30 minutes at 4°C with gentle shaking. The resulted suspension is centrifuged at 100,000 g for 30 minutes. The supernatant is collected and mixed with 1 mL of NTA-nickel resin equilibrated with HEPES-buffered saline. The mixture is incubated for 1 hour at 4°C with gentle shaking. The resin is then rinsed in buffer and spun at 100 × g for 1 minutes. The supernatant is removed and the beads resuspended in a washing buffer containing 100 mM imidazole-HCl, pH 7.5. The rinsing process is repeated three times. The final purified fraction is eluted in 2 mL of buffer with 250 mM imidazole-HCl. To remove imidazole, the eluate is dialyzed overnight against 1 L of HEPES buffered saline at 4°C with gentle stirring.
- **Protein immobilization on the glass surface:** To immobilize protein on the surface of the glass, the glass rod is cleaned with a plasma cleaner (plasma cleaner Zepto, Diener Electronic, Ebhausen, Germany) using 80% output for 5 minute at 0.3 mbar under oxygen gas. The cleaned glass rods are rinsed in analytical-grade ethanol (>99.9% purity, Sigma) and incubated in 5% 3-(aminopropyl) triethoxysilane in ethanol for 10 minutes at room temperature. The treated glass rods are then rinsed with ethanol and baked at 80°C for 30 minutes. To deprotonate the functionalized amines, the glass rods are soaked in 20 mM Na_2HPO_4 overnight. To functionalize them with the maleimide groups, the glass rods are incubated in 5 mM NHS-PEG$_{(27)}$-maleimide (Polypure, Oslo, Norway) in a phosphate-buffered saline (PBS: 10 mM Na_2HPO_4, 1.76 mM KH_2PO_4, 137 mM NaCl, 2.7 mM KCl) for 1 hour at room temperature. The glass rods are then incubated in 20 mM coenzyme A trilithium salt (Sigma) for 1 hour at room temperature. After rinsed with Milli-Q water, the glass rods are functionalized with the protein by incubation with 100 μg/mL ybbR-4 x spectrin-XMod-dockerin in

the presence of 1 μM 4′-phosphopantetheinyl transferase (SFP synthase) in 25 mM Tris-HCl, pH 7.2, 75 mM NaCl, 1 mM CaCl$_2$, and 1 mM MgCl$_2$ for 1 hour at room temperature. The protein-covered glass rods are then immediately used for measurements.

- **Tip preparation:** HS-AFM cantilevers (BL-AC10DS, Olympus, Japan) are used for HS-AFM-FS experiments. They are initially cleaned in the plasma cleaner and functionalized with coenzyme A. The cantilevers are then functionalized with protein by incubation in 100 μg/mL ybbRcohesin-III in the presence of 1 μM SFP synthase in 25 mM Tris-HCl, pH 7.2, 75 mM NaCl, 1 mM CaCl$_2$, and 1 mM MgCl$_2$ for 1 hour at room temperature. After preparation, the cantilevers are immediately used for measurements.

5.3.3.1.2.1 HS-AFM-FS Experiments In the HS-AFM-FS experiments, the deflection sensitivity and spring constant of the cantilevers are determined before every experiment using the thermal fluctuations method.[57] The used cantilevers (BL-AC10DS) have typical resonance frequencies of ~1.3 and ~0.6 MHz in air and liquid, respectively. The HS-AFM-FS setup is controlled with a home-built software package using a multichannel analog-to-digital converter with a maximum acquisition rate of 20 Mega samples per second in each channel, which drives piezo displacement and acquires cantilever deflection (LabView programming, PXI-5122 card, National Instruments, USA). The spring constants of the cantilevers are determined using the Sader method by applying the correction for higher modes corresponding to rectangular cantilevers.[58] The sensitivity of the photodiode is extracted from the thermal noise spectrum using the calibrated spring constant. Force spectroscopy measurements are performed in a buffer containing 25 mM Tris-HCl, pH 7.5, 75 mM NaCl, and 2 mM CaCl$_2$ by approaching the sample surface to the cantilever at ~1 μm/s velocity. The contact was maintained for 1 s before the retraction of the sample surface in velocities ranging from 0.1 to 10^4 μm/s.

5.3.3.1.2.2 Typical results Given the capability of HS-AFM-FS, Takahashi et al. used HS-AFM-FS to analyze spectrin unfolding, the data was comparable to SMDS and thus allowed direct comparison of experimental and simulated unfolding for the characterization of spectrin mechanics over a wide dynamic range, and deduce atomistic interpretations of the forced unfolding from the comparison with SMDS. The results demonstrated that spectrin is a viscoelastic molecular force buffer, occasionally acting as a soft spring at short extensions, followed by a long-distance viscous response dominated by the unwinding of α-helices, which is diagramed in Figure 5.10. Other interesting results from pulling proteins could be found in references.[59–61] Please note that this is a method of pulling proteins at constant velocity. In other different applications, it is also possible to pull proteins at constant force, which is termed force-clamp measurements.[62,63]

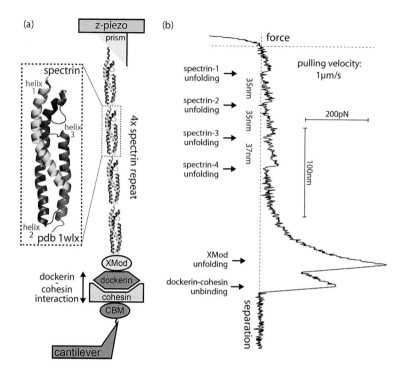

FIGURE 5.10 HS-AFM-FS setup for spectrin unfolding experiments. (a) Schematic illustration of the HS-AFM-FS setup for spectrin unfolding: spectrin (PDB: 1WLX) concatamers of four repeats are immobilized to the glass substrate via a covalent bond. The spectrin concatemers are chimerized with Xmod and dockerin domains. On the HS-FS cantilever tip, cohesin, the binding partner of dockerin, is also immobilized covalently. (b) Force–extension curve showing four unfolding peaks. Since the cohesin–dockerin pair binds strongly (>400 pN unbinding force), the force distance curves are selected for the Xmod unfolding and cohesin–dockerin unbinding peak at a distance of ~200 nm from the surface corresponding to the unfolding of the four spectrin repeats (with each spectrin repeat 129 amino acids long and an extension of 0.38 nm per amino acid; a full extension of ~49 nm per spectrin repeat is expected). (Reprinted with permission from Hirohide, T. et al., *ACS Nano*, 12, 2719–2727, 2018. Copyright 2018 American Chemical Society.)

5.3.3.2 HS-AFM Experiments of Integral Membrane Proteins

Integral membrane proteins, also known as transmembrane proteins, are embedded within the bilayer membranes of a cell, and are generally responsible for signal transduction, compound transport or channeling, etc. Importantly, most transmembrane proteins function with a conformational change as indicated by X-ray structures. However, a direct visualization of the conformational changes of a transmembrane protein remained elusive in static structures, nor can dynamic parameters be deduced from them. HS-AFM provides a unique method to probe membrane proteins in native-like lipid membranes. With the recent developments, it enables us to control the environment and resolve the details of conformational change of membrane proteins in real time.

Step-by-step sample preparation protocol

- **Proteins purification:** The purification protocol of transmembrane proteins depends strongly on the specific protein under investigation. Some general steps are typically involved and detailed here in a protocol for purifying the prokaryotic glutamate transporter, Glt_{Ph}.[27,64] Briefly, the protein is expressed in *Escherichia coli* DH10b strain as a C-terminal fusion with a thrombin cleavage site and a $(His)_8$ tag. The isolated crude membranes are extracted in buffer A, containing 20 mM Hepes/Tris, pH 7.4, 200 mM NaCl, 0.1 mM L-aspartate, and 40 mM n-dodecyl β-D-maltopyranoside (DDM) for 2 hour at 4°C (the detergent is a crucial parameter and must be adapted for each protein to assure its stability during membrane extraction). Solubilized transporters are applied to immobilized metal affinity resin in buffer A in the presence of 1 mM DDM. The resin is washed in the same buffer supplemented with 40 mM imidazole, and the protein is eluted with 250 mM imidazole. The $(His)_8$ tag is removed by thrombin digestion overnight, and protein is further purified by size-exclusion chromatography in the following buffer: 10 mM Hepes, pH 7.4, 100 mM NaCl, 0.1 mM L-aspartate, 0.4 mM DDM. The protein is concentrated to ~7 mg/mL, flash-frozen, and stored at −80°C. Protein concentration is determined by absorbance at 280 nm using an extinction coefficient of 57,400 $M^{-1} \cdot cm^{-1}$ (per monomer).

- **Protein reconstitution:** Although different methods have been proposed to insert integral membrane proteins into liposomes, detergent removal through the addition of Bio-Beads or dialysis from lipid-protein-detergent mixtures is the most successful and experimentally verified way to reconstitute integral membrane proteins into proteoliposomes and obtain 2D crystals for HS-AFM analysis.[65–67] In accord with the above description, pure glutamate transport protein (as judged by SDS/PAGE) at 0.44 mg/mL in 20 mM Tris-HCl (pH 7.5), 300 mM NaCl, 2 mM Tris(2-carboxyethyl)phosphine (TCEP), and 0.02% dodecylmaltoside (DDM) was mixed with a lipid mixture of 1,2-dioleoyl-sn-glycero-3-phosphocholine (DOPC)/1,2-dioleoyl-sn-glycero-3-phosphoethanolamine (DOPE)/1,2-dioleoyl-snglycero-3-phospho-L-serine (DOPS) in 8:1:1 weight ratio or *Escherichia coli* lipids at a lipid-to-protein ratio of ~1 (wt/wt) and complemented with a buffer containing 10 mM Hepes (pH 8.0), 200 mM KCl, 2 mM NaN_3, and 0.025% DDM or 10 mM Na-acetate (pH 4.5), 200 mM KCl, 2 mM NaN_3, and 0.025% DDM to a final protein concentration of 0.3 mg/mL. The protein/lipid/detergent mixture is allowed to equilibrate for 2 hour before Bio-Beads (Bio-Rad, CA, USA) are added for detergent removal overnight at room temperature. Progress of reconstitution is followed by negative-stain EM. Once intact, densely packed vesicles are detected by EM, the Bio-Beads are removed from the sample.

- **Sample preparation:** ~2 μL of reconstituted glutamate transporter vesicles are injected onto a 1-mm-diameter freshly cleaved mica. HS-AFM sample support covered with adsorption buffer containing 100 mM Mg^{2+}. The vesicles are allowed to adsorb for 15 minutes, followed by ×10 wash with the imaging buffer.

5.3.3.2.1   HS-AFM Experiments

Short cantilevers with a nominal spring constant of 0.15 N/m (Type: USC-F1.2-k0.15, NanoWorld, Switzerland), resonance frequency of 0.6 MHz and quality factor of ~1 in liquid are used. Typical image acquisition parameters are: image size of 80 × 80 nm, frame size of 200 × 200 pixels, image acquisition rate of 1 frame per second, free amplitude of 1 nm, and set point amplitude of 0.9 nm. A constant-pressure and constant-flow pump (Harvard Instruments) is connected through silicon tubes to the fluid cell pool of the cantilever holder of the HS-AFM microscope for injecting and extracting buffers during HS-AFM observation. The buffers are exchanged during HS-AFM observation. A tiny air bubble in the injecting buffer tube at the interface to the fluid cell initially separates the injecting buffer from the fluid cell pool and prevents fluid mixing in the fluid cell before active buffer exchange. The buffer exchange flow rate is fixed at 5, 10, or 15 μL·min^{-1}. Note that higher flow rates may cause perturbation induced by the buffer flow.

5.3.3.2.2   Typical Results

Transporter proteins are key integral membrane proteins in many physiological processes. In the nervous system the glutamate transporter recovers the main excitatory neurotransmitter, glutamate, from the synaptic cleft. Their dysfunction is associated with many neurological diseases, such as epilepsy, Alzheimer's disease, and amyotrophic lateral sclerosis. The glutamate transporter crystal structures of a prokaryotic homolog of archaebacterium *Pyrococcus horikoshii* is solved in both the outward- and inward-facing conformations of the sodium/aspartate symporter termed Glt$_{Ph}$ to reveal the molecular basis of the transport mechanism.[64] HS-AFM imaging revealed directly the "elevator" typ transport mechanism of Glt$_{Ph}$, where each transporter domain moves vertically across the membrane, as shown in Figure 5.11. HS-AFM also revealed non-cooperativity between the subunits in the trimeric Glt$_{Ph}$. In order to show how the environmental conditions, the ligand, co-factor or inhibitor TBOA (DL-threo-β-Benzyloxyaspartic acid) concentrations, change Glt$_{Ph}$ activity and its transport dynamics. Real-time buffer transition experiments are performed using HS-AFM coupled to a buffer-exchange system. By exchanging the measuring buffer (using blue, red tubes in Figure 5.5) from a substrate-free condition to a NaCl/aspartate-rich environment, or by the injection of the inhibitor TBOA, it was shown that there is a rather gradual silencing of Glt$_{Ph}$ activity upon the addition of Na$^+$ with an approximate half-maximal activity in the presence of NaCl at a few hundred mM. In case of TBOA addition, the vast majority of transporters are quiescent at ~100 μM TBOA.[27]

In the Glt$_{Ph}$ example, the implemented and optimized buffer-exchange system is used to gradually change the substrate/inhibitor concentration while recording high-resolution movies. However, to test the reversibility of a biological process, an additional syringe with a third buffer condition, identical to the first, is added to the buffer exchange device (see Figure 5.5).

Using this multi-tube pumping system, Ruan et al.[28] investigated the GLIC channel (*Gloeobacter violaceus* pentameric ligand-gated ion channel homologue), which is a prokaryotic homologue of the Cys-loop pentameric receptor ligand-gated ion channel family. It is a proton-gated, cation-selective channel, as shown in Figure 5.12.

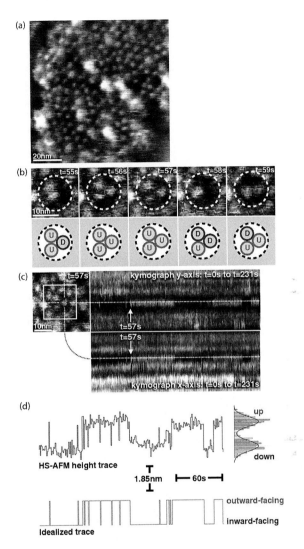

FIGURE 5.11 Direct visualization of Glt$_{Ph}$ "elevator" domain movements by HS-AFM. (a) A frame from a typical HS-AFM movie of a membrane containing densely packed glutamate transporter trimers. Full color scale is 8 nm. (b) Sequential frames displaying the conformational dynamics of a Glt$_{Ph}$ trimer under substrate-free conditions. The trimer structure is only discernible to the eye in frames where all three subunits are outward facing (U, $t = 56$ s, $t = 57$ s). (c, Upper Left) Frame $t = 57$ s from image series shown in b (white square). (c, Right) Kymographs of the Y (Upper) and X (Lower) section profiles across a transport domain (dashed lines). Transition from down to up position at approximately $t = 57$ s is highlighted in the kymographs. Full color scale in b and c is 3 nm. (d, Upper) Height trace (average from the X and Y kymographs) as a function of time, and height value distribution histogram (Right). (d, Lower) Idealized trace following assignment of up and down states to outward- and inward-facing conformations of the Glt$_{Ph}$ elevator domain, respectively. (Reprinted with permission from Ruan, Y. et al., 2017. Direct visualization of glutamate transporter elevator mechanism by high-speed AFM. *Proc. Nat. Acad Sci USA*, 114, 1584. Copyright 2017 National Academy of Sciences, USA.)

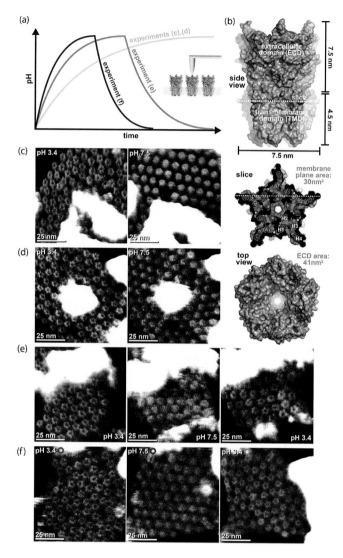

FIGURE 5.12 HS-AFM structural titration experiments of GLIC channels. (a) Schematic representation of the experimental procedure. (Inset) During HS-AFM imaging, the pH is gradually changed in the fluid chamber of 150 μL in volume through a buffer exchange microfluidic device. The flow rate is adjusted to 5 μL·min⁻¹ (experiments shown in C and D), 10 μL·min⁻¹ (experiment shown in E), and 15 μL·min⁻¹ (experiment shown in F). In the experiments shown in E and F, the pH was reversed to observe reversibility of assembly and structural changes. (b) Analysis of the GLIC architecture. A side view (Top), membrane slice close to the extracellular surface (Middle), and top view (Bottom) are shown. The dimensions of the molecule are indicated. High-resolution images (time averages over 5 s) of the initial pH 3.4 state (c and d, Left; open channel) and pH 7.5 state (c and d, Right; closed channel), as well as the reversed pH 3.4 state (e and f, Right; open channel) are shown. (Adapted from Ruan, Y. et al., 2018. Structural titration of receptor ion channel GLIC gating by HS-AFM, *Proc. Nat. Acad Sci USA*, 115, 10333. Copyright 2018 National Academy of Sciences.)

The gating mechanisms of this receptor channel family is inferred from comparing X-ray structures,[68] but remain elusive on one protein and certainly on single molecules in real time. HS-AFM is used to observe that the GLIC pentamers are non-ordered in the membrane at pH 3.4 buffer condition, appear as pentagons consisting of five subunits with a top-ring diameter of ~3 nm and a central cavity into which the HS-AFM tip could penetrate by ~1.5 nm. Significant conformational changes were observed upon operating the reversible buffer exchanging system while taking HS-AFM movies, exposing the channels to pH 3.4, then pH 7.5 and again to pH 3.4. While the molecules were imaged at a frame rate of 1 s^{-1} at a submolecular resolution, a pH 7.5 buffer slowly was titrated into the fluid chamber. During this process, the channels underwent stunning dynamics. Molecules wiggle around their position; the ECDs narrow toward to the central axis; lateral diffusion increases; and finally, the channels stabilize with apparently tightened ECDs, which only rarely allow the detection of a central cavity, in a hexagonal lattice. During a second titration, back to the initial pH 3.4 buffer, using a third syringe allowed to image the recovery of the open-ECD conformation and lateral rearrangements to a fully reversible reconfiguration to the initial state.

5.3.3.3 HS-AFM Experiments on Cytosolic Membrane Remodeling Proteins

During membrane trafficking, a myriad of proteins act in concert to promote the formation of membrane carriers through membrane remodeling. Previous studies have shown that membrane remodeling proteins perform their functions through several modes. ESCRT-III (Endosomal Sorting Complex Required for Transport) has been implicated in the formation of Intralumenal Vesicles (ILVs) during biogenesis of Multi-Vesicular Bodies (MVBs) by genetic[69] and biochemical assays.[70,71] This budding process has a topology opposite to the membrane invaginations occurring during endocytosis and membrane traffic at the endoplasmic reticulum or Golgi. In MVBs, the limiting membrane is pushed outwards from the cytoplasm instead of curving inwards. ESCRT-III has been proposed to play a role in membrane deformation[72] and fission of ILVs.[70] However, it is unclear how ESCRT-III deforms lipid membranes. Because of their polymerization abilities, ESCRT-III proteins (Vps20, Snf7, Vps2, Vps24) have been proposed to generate membrane curvature by scaffolding.[73,74]

Step-by-step sample preparation protocol

- **Proteins purification:** Snf7 (addgene plasmid n°21492), Escrt II (addgene plasmid n°17633) and Vps20 (addgene plasmid n°21490) were purified as previously described.[75,76] Snf7 stock solution was 5 µM in Hepes 20 mM pH 8 NaCl 100 mM. This concentration was used as a reference for batches of labeled molecules.
- **Giant unilamellar vesicles (GUVs) preparation:** GUVs were prepared by electroswelling[77] using a mixture of 1,2-dioleoyl-sn-glycero-3-phospho-choline (DOPC) and 1,2-dioleoyl-sn-glycero-3-phospho-L-serine (DOPS), both of which were purchased from Avanti Polar Lipids (Alabaster, AL, USA). 20 µL of 1 mg/mL lipids (DOPC: DOPS, 6:4, mol:mol) was dissolved in chloroform, deposited on two glass plates coated with indium tin oxide (ITO) (70–100 Ω resistivity, Sigma-Aldrich) and placed in a vacuum drying

oven for at least 60 minutes for complete solvent evaporation. A U-shape rubber piece of ~1 mm thickness was sandwiched on the borders between the two ITO slides. The formed chamber was filled with 400 µL of 500 mM sucrose solution (osmolarity adjusted to that of the outside buffer solution, around 500 mOsm) and exposed to 1 V AC-current (10 Hz sinusoidal for 1 hour) at room temperature. The resulting suspension was collected in a tube and used within the next days for experiments.

- The buffer used for all experiments is composed of NaCl 200 mM, Tris-HCl 20 mM pH 6.8, and $MgCl_2$ 1 mM.
- **Sample preparation:** GUVs prepared as described above, were adsorbed to the mica support followed by protein addition. 0.5 µL of the GUVs was deposited onto freshly cleaved mica supports pre-incubated with 2 µL of adsorption buffer (220 mM NaCl, 10 mM Hepes, 2 mM $MgCl_2$, pH 7.4). After 15′ incubation, samples were gently washed with the measurement buffer (10 mM Tris-HCl, 150 mM KCl, pH 7.4). Supported lipid bilayers were first imaged to assess the quality of the lipid bilayer preparation before Snf7 was injected into the fluid cell to a concentration of ~1 µM. Formation of Snf7 assemblies was observable ~30 minutes after Snf7 injection.

5.3.3.3.1   HS-AFM Experiments

HS-AFM was performed in a HS-AFM 1.0 (RIBM, Tsukuba, Japan) setup equipped with 8 µm long cantilevers with a nominal spring constant $k = 0.15$ N/m and a resonance frequency $f(r) = 1200$ kHz in air (NanoWorld, Neuchâtel, Switzerland), both of which were calibrated before each experiment. HS-AFM image acquisition was performed using a dynamic feedback circuit at the highest possible speed. Images were obtained at a pixel sampling ranging from 200 by 200 to 500 by 500 pixels on scanned areas ranging from 200 to 1500 nm^2. Under non-destructive force conditions, movies at frame rates ranging from 1 to 5 frames per second could be acquired. HS-AFM movies were stabilized for movie acquisition piezo drift in post-acquisition treatment as described.[78]

5.3.3.3.2   Typical Results

In Chiaruttini et al.,[79] HS-AFM has been used to study membrane remodeling processes, particularly the fission machinery ESCRT-III. ESCRT-III (Endosomal Sorting Complex Required for Transport) is needed for lipid membrane remodeling in many cellular processes, from abscission to viral budding and formation of late endosomes. However, how ESCRT-III polymerization generates membrane curvature remains debated. In this work it was shown that Snf7, the main component of ESCRT-III, polymerized into spirals at the surface of supported lipid bilayers (Figure 5.13).

HS-AFM movies revealed the Snf7 complex formation and its dynamics from filament to the maturated spiral assembly around the membrane constriction site. The observed inter-filament dynamics forges a basis for a mechanistic explanation of how this protein machinery creates tension for membrane fission. Furthermore, this study showed that Snf7 assemblies compress the inner diameter during maturation,

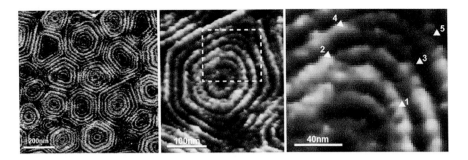

FIGURE 5.13 High-resolution AFM topographic image of a membrane area containing Snf7 patches. A large number of patches are shown in the panel on the left and one of them is enlarged in the center panel. The square drawn in this enlarged view is zoomed in and showed to the right, where it reveals splits and variability of Snf7 structures. (Reproduced from Chiaruttini, N. et al., *Cell*, 163, 866–879, 2015 [13].)

which constitutes a direct evidence of force generation during the assembly process. When the spirals were disrupted with the AFM cantilever, the broken polymers spontaneously rearranged to form smaller rings, suggesting a preferred radius curvature of the Snf7 polymer of ~25 nm. Based on these observations, it was speculated that Snf7 spirals could function as spiral springs. From measurements of the polymerization energy and the rigidity of Snf7 filaments, it was concluded that the filaments were deformed while growing in a confined area.

Furthermore, the dynamic data also suggested that the elastic expansion of compressed Snf7 spirals could stretch the lipids they are bound to, generating an area difference between the membrane leaflets and thus inducing curvature. This spring-like activity might be the driving force by which ESCRT-III could mediate membrane remodeling.

5.4 PERSPECTIVES

HS-AFM provides direct visualization of structural and dynamic information on membrane proteins with lateral resolutions of ~1 nm lateral and a vertical resolution of ~0.1 nm. It requires neither labeling nor complex data interpretation. Together with the structural analysis of membrane proteins by X-ray crystallography or cryo-electron microscopy, it helps us to gain further insights into the physiological function of membrane proteins, as well as into the dynamics of protein–protein interactions. The recent developments, e.g., buffer exchange system, temperature control, combination with light microscopy, etc., present novel possibilities for HS-AFM to study membrane proteins in varying environmental conditions, and to help us better understand the integrated function of membrane proteins.

As mentioned above, recent developments extended HS-AFM, to so-called high-speed AFM height spectroscopy (HS-AFM-HS),[80] whereby it is used to monitor the sensing of a HS-AFM tip at a fixed position to directly detect the motions of unlabeled molecules underneath. By these means, it is possible to achieve Angstrom-level spatial and microsecond-scale temporal resolutions. Briefly, the AFM tip is held at a fixed $x-y$ position and monitors the height fluctuations under the tip in z-direction. The temporal resolution is only limited by the feedback speed and in the best case, when the oscillation amplitude is recorded, the response time of the cantilever.

Apart from what is discussed so far, wide-area/fast scanners have recently been developed.[81] In combination with tip-scan (not sample-scan) HS-AFM, the observation of dynamic changes in living cells becomes possible. Tip-scanning HS-AFM, though somewhat inferior in performance compared to sample-scanning HS-AFM, can potentially be used in combination to other techniques such as super-resolution optical microscopy. For eukaryotic cells, which can be extremely soft, High-Speed Scanning Ion Conductance Microscopy, which probes the sample in non-contact mode, might be a promising avenue. HS-AFM combined with optical tweezers might allow the visualization of biomolecules under the application of an external force.[82]

The field of HS-AFM is under continuous development. We foresee that it will expand, together or in combination with other techniques, in the near future, and will exert high impact in biophysics. Furthermore, with the recent breakthrough in cryo electron microscopy, structures of membrane proteins are now solved rather routinely. HS-AFM may contribute information about protein-protein interactions, real-time conformational changes, and kinetics of these molecules in action.

ACKNOWLEDGMENTS

Y.R. acknowledges support from National Nature Science Foundation of China Grants, NSFC [61805213] and the Fundamental Research Funds for the Provincial Universities of Zhejiang (Zhejiang Provincial Natural Science Foundation of China under Grant No. [LGF20C050001]).

L.R-M. acknowledges support from the French National Institute of Health and Medical Research (Inserm) and by a grant overseen by the French National Research Agency (ANR) as part of the "Investments d'Avenir" Programme (I-SITE ULNE/ ANR-16-IDEX-0004 ULNE).

S.S acknowledges support from National Institutes of Health (Grant numbers: DP1AT010874 from NCCIH and RO1NS110790 from NINDS).

REFERENCES

1. Binnig, G., Quate, C. F. & Gerber, C. Atomic force microscope. *Physical Review Letters* **56**, 930–933, doi:10.1103/PhysRevLett.56.930 (1986).
2. Binnig, G., Gerber, C., Stoll, E., Albrecht, T. R. & Quate, C. F. Atomic resolution with atomic force microscope. *Europhysics Letters* **3**, 1281 (1987).
3. Drake, B. et al. Imaging crystals, polymers, and processes in water with the atomic force microscope. *Science* **243**, 1586–1589 (1989).
4. Engel, A. Biological applications of scanning probe microscopes. *Annual Review of Biophysics and Biophysical Chemistry* **20**, 79–108 (1991).
5. Müller, D. J., Schabert, F. A., Büldt, G. & Engel, A. Imaging purple membranes in aqueous solutions at sub-nanometer resolution by atomic force microscopy. *Biophysical Journal* **68**, 1681–1686 (1995).

6. Fechner, P. et al. Structural information, resolution, and noise in high-resolution atomic force microscopy topographs. *Biophysical Journal* **96**, 3822–3831 (2009).

7. Müller, D. J. et al. Single-molecule studies of membrane proteins. *Current Opinion in Structural Biology* **16**, 489–495 (2006).

8. Ando, T. et al. A high-speed atomic force microscope for studying biological macro-molecules. *Proceedings of the National Academy of Sciences of the United States of America* **98**, 12468–12472 (2001).

9. Viani, M. B., Schaffer, T. E., Paloczi, G. T. & Pietrasanta, L. I. Fast imaging and fast force spectroscopy of single biopolymers with a new atomic force microscope designed for small cantilevers. *Review of Scientific Instruments* **70**, 4300–4303 (1999).

10. Viani, M. B. et al. Probing protein-protein interactions in real time. *Nature Structural & Molecular Biology* **7**, 644–647 (2000).

11. Schaffer, T. E., Cleveland, J. P., Ohnesorge, F., Walters, D. A. & Hansma, P. K. Studies of vibrating atomic force microscope cantilevers in liquid. *Journal of Applied Physics* **80**, 3622–3627 (1996).

12. Kindt, J. H., Fantner, G. E., Cutroni, J. A. & Hansma, P. K. Rigid design of fast scan-ning probe microscopes using finite element analysis. *Ultramicroscopy* **100**, 259–265 (2004).

13. Kodera, N., Yamashita, H. & Ando, T. Active damping of the scanner for high-speed atomic force microscopy. *Review of Scientific Instruments* **76**, 4300 (2005).

14. Fukuma, T., Okazaki, Y., Kodera, N., Uchihashi, T. & Ando, T. High resonance fre-quency force microscope scanner using inertia balance support. *Applied Physics Letters* **92**, 930 (2008).

15. Kodera, N., Sakashita, M. & Ando, T. Dynamic proportional-integral-differential con-troller for high-speed atomic force microscopy. *Review of Scientific Instruments* **77**, 1 (2006).

16. Butt, H. J. et al. Scan speed limit in atomic force microscopy. *Journal of Microscopy* **169**, 75–84, doi:10.1111/j.1365-2818.1993.tb03280.x (1993).

17. Walters, D. A. et al. Short cantilevers for atomic force microscopy. *Review of Scientific Instruments* **67**, 3583–3590 (1996).

18. Viani, M. B. et al. Fast imaging and fast force spectroscopy of single biopolymers with a new atomic force microscope designed for small cantilevers. *Review of Scientific Instruments* **70**, 4300–4303 (1999).

19. Manalis, S. R., Minne, S. C. & Quate, C. F. Atomic force microscopy for high speed imaging using cantilevers with an integrated actuator and sensor. *Applied Physics Letters* **68**, 871–873 (1996).

20. Lutwyche, M. et al. 5 × 5 2D AFM cantilever arrays a first step towards a Terabit stor-age device. *Sensors and Actuators* **73**, 89–94(86) (1999).

21. Ando, T., Uchihashi, T. & Kodera, N. High-speed AFM and applications to biomolecu-lar systems. *Annual Review of Biophysics* **42**, 393–414 (2013).

22. Ando, T., Uchihashi, T. & Fukuma, T. High-speed atomic force microscopy for nano-visualization of dynamic biomolecular processes. *Progress in Surface Science* **83**, 337–437 (2008).

23. Ando, T. et al. High-speed atomic force microscopy for studying the dynamic behavior of protein molecules at work. *e-Journal of Surface Science and Nanotechnology* **3**, 384–392 (2005).

24. Albrecht, T. R., Grütter, P., Horne, D. & Rugar, D. Frequency modulation detection using high-Q cantilevers for enhanced force microscope sensitivity. *Journal of Applied Physics* **69**, 668–673 (1991).

25. Miyagi, A., Chipot, C., Rangl, M. & Scheuring, S. High-speed atomic force micros-copy shows that annexin V stabilizes membranes on the second timescale. *Nature Nanotechnology* **11**, 783 (2016).

26. Rangl, M. et al. Real-time visualization of conformational changes within single MloK1 cyclic nucleotide-modulated channels. *Nature Communications* **7**, 12789 (2016).

27. Ruan, Y. et al. Direct visualization of glutamate transporter elevator mechanism by high-speed AFM. *Proceedings of the National Academy of Sciences of the United States of America* **114**, 1584 (2017).

28. Ruan, Y. et al. Structural titration of receptor ion channel GLIC gating by HS-AFM. *Proceedings of the National Academy of Sciences of the United States of America* **115**, 10333–10338, doi:10.1073/pnas.1805621115 (2018).

29. Enders, O., Ngezahayo, A., Wiechmann, M., Leisten, F. & H-A, K. Structural calorimetry of main transition of supported DMPC bilayers by temperature-controlled AFM. *Biophysical Journal* **87**, 2522–2531 (2004).

30. Takahashi, H., Miyagi, A., Redondo-Morata, L. & Scheuring, S. Temperature-controlled high-speed AFM: Real-time observation of ripple phase transitions. *Small* **12**, 6106–6113 (2016).

31. Wouter, E. et al. The capsaicin receptor TRPV1 is a crucial mediator of the noxious effects of mustard oil. *Current Biology* **21**, 316–321 (2011).

32. Colom, A., Casuso, I., Rico, F. & Scheuring, S. A hybrid high-speed atomic force: Optical microscope for visualizing single membrane proteins on eukaryotic cells. *Nature Communications* **4**, 2155, doi:10.1038/ncomms3155 (2013).

33. Yamashita, H. et al. Single-molecule imaging on living bacterial cell surface by high-speed AFM. *Journal of Molecular Biology* **422**, 300–309 (2012).

34. Pastré, D., Iwamoto, H., Liu, J., Szabo, G. & Shao, Z. Characterization of AC mode scanning ion-conductance microscopy. *Ultramicroscopy* **90**, 13–19 (2001).

35. Happel, P., Hoffmann, G., Mann, S. A. & Dietzel, I. D. Monitoring cell movements and volume changes with pulse-mode scanning ion conductance microscopy. *Journal of Microscopy* **212**, 144–151 (2010).

36. Novak, P. et al. Corrigendum: Nanoscale live-cell imaging using hopping probe ion conductance microscopy. *Nature Methods* **6**, 279 (2009).

37. Miyagi, A. & Scheuring, S. Automated force controller for amplitude modulation atomic force microscopy. *Review of Scientific Instruments* **87**, 930 (2016).

38. Miyagi, A. & Scheuring, S. A novel phase-shift-based amplitude detector for a high-speed atomic force microscope. *The Review of Scientific Instruments* **89**, 083704, doi:10.1063/1.5038095 (2018).

39. Chen, C.-C., Zhou, Y. & Baker, L. A. Scanning ion conductance microscopy. *Annual Review of Analytical Chemistry* **5**, 207–228, doi:10.1146/annurev-anchem-062011-143203 (2012).

40. Hansma, P. K., Drake, B., Marti, O., Gould, S. A. & Prater, C. B. The scanning ion-conductance microscope. *Science* **243**, 641–643 (1989).

41. Ying, L. et al. The scanned nanopipette: A new tool for high resolution bioimaging and controlled deposition of biomolecules. *Physical Chemistry Chemical Physics* **7**, 2859–2866, doi:10.1039/B506743J (2005).

42. Shevchuk, A. I. et al. Imaging proteins in membranes of living cells by high-resolution scanning ion conductance microscopy. *Angewandte Chemie* **45**, 2212–2216 (2006).

43. Miragoli, M. et al. Scanning ion conductance microscopy: A convergent high-resolution technology for multi-parametric analysis of living cardiovascular cells. *Journal of the Royal Society, Interface* **8**, 913–925, doi:10.1098/rsif.2010.0597 (2011).

44. Kemiktarak, U., Ndukum, T., Schwab, K. C. & Ekinci, K. L. Radio-frequency scanning tunnelling microscopy. *Nature* **450**, 85, doi:10.1038/nature06238 (2007).

45. Scheuring, S. https://bio-afm-lab.com/ (accessed January 20, 2020).

46. *Biological Membrane*, https://en.wikipedia.org/wiki/Biological_membrane (accessed December 23, 2019).

47. Mathews, C. K., Holde, K. E. V., Appling, D. R. & Anthony-Cahill, S. J. *Biochemistry.* 4th edn. (Toronto, ON: Pearson Education Canada, 2012).

48. Cerne, K., Kristan, K. C., Budihna, M. V. & Stanovnik, L. Mechanisms of changes in coronary arterial tone induced by bee venom toxins. *Toxicon: Official Journal of the International Society on Toxinology* **56**, 305–312, doi:10.1016/j.toxicon.2010.03.010 (2010).

49. Mingeotleclercq, M. P., Deleu, M., Brasseur, R. & Dufrêne, Y. F. Atomic force microscopy of supported lipid bilayers. *Nature Protocols* **3**, 1654–1659 (2008).

50. Ruan, Y., Rezelj, S., Bedina, Z. A., Anderluh, G. & Scheuring, S. Listeriolysin O membrane damaging activity involves arc formation and lineaction: Implication for *Listeria monocytogenes* escape from phagocytic vacuole. *Plos Pathogens* **12**, e1005597 (2016).

51. Munguira, I. et al. Glasslike membrane protein diffusion in a crowded membrane. *ACS Nano* **10**, 2584–2590, doi:10.1021/acsnano.5b07595 (2016).

52. Rief, M., Gautel, M., Oesterhelt, F., Fernandez, J. M. & Gaub, H.E. Reversible unfolding of individual titin immunoglobulin domains by AFM. *Science* **276**, 1109–1112 (1997).

53. Florin, E. L., Moy, V. T. & Gaub, H. E. Adhesion forces between individual ligand-receptor pairs. *Science* **264**, 415–417 (1994).

54. Takahashi, H., Rico, F., Chipot, C. & Scheuring, S. α-Helix unwinding as force buffer in spectrins. *ACS Nano* **12**, 2719–2727 (2018).

55. Soncini, M. et al. Mechanical response and conformational changes of alpha-actinin domains during unfolding: A molecular dynamics study. *Biomechanics & Modeling in Mechanobiology* **6**, 399–407 (2007).

56. Isralewitz, B., Gao, M. & Schulten, K. Steered molecular dynamics and mechanical functions of proteins. *Current Opinion in Structural Biology* **11**, 224–230 (2001).

57. Hutter, J. L. & Bechhoefer, J. Calibration of atomic: Force microscope tips. *Review of Scientific Instruments* **64**, 1868–1873, doi:10.1063/1.1143970 (1993).

58. Sader, J. E., Chon, J. W. M. & Mulvaney, P. Calibration of rectangular atomic force microscope cantilevers. *Review of Scientific Instruments* **70**, 3967–3969 (1999).

59. Felix, R., Laura, G., Ignacio, C., Manel, P. V. & Simon, S. High-speed force spectroscopy unfolds titin at the velocity of molecular dynamics simulations. *Science* **342**, 741–743 (2013).

60. Sumbul, F., Marchesi, A., Takahashi, H., Scheuring, S. & Rico, F. High-speed force spectroscopy for single protein unfolding. *Methods in Molecular Biology* **1814**, 243 (2018).

61. Lu, H., Isralewitz, B., Krammer, A., Vogel, V. & Schulten, K. Unfolding of titin immunoglobulin domains by steered molecular dynamics simulation. *Biophysical Journal* **75**, 662 (1998).

62. Hughes, M. L. & Dougan, L. The physics of pulling polyproteins: A review of single molecule force spectroscopy using the AFM to study protein unfolding. *Reports on Progress in Physics Physical Society* **79**, 076601 (2016).

63. Schönfelder, J., Perez-Jimenez, R. & Muñoz, V. A simple two-state protein unfolds mechanically via multiple heterogeneous pathways at single-molecule resolution. *Nature Communications* **7**, 11777, doi:10.1038/ncomms11777 (2016).

64. Dinesh, Y., Olga, B., Yan, J. & Eric, G. Structure of a glutamate transporter homologue from Pyrococcus horikoshii. *Nature* **431**, 811–818 (2004).

65. Rigaud, J. L., Levy, D., Mosser, G. & Lambert, O. Detergent removal by non-polar polystyrene beads. *European Biophysics Journal* **27**, 305–319 (1998).

66. Rigaud, J. L. et al. Bio-beads: An efficient strategy for two-dimensional crystallization of membrane proteins. *Journal of Structural Biology* **118**, 226 (1997).

67. Rigaud, J. L. & Lévy, D. Reconstitution of membrane proteins into liposomes. *Methods Enzymology* **372**, 65–86 (2003).

68. Nicolas, B. et al. X-ray structure of a pentameric ligand-gated ion channel in an apparently open conformation. *Nature* **457**, 111–114 (2009).

69. Babst, M., Katzmann, D. J., Estepa-Sabal, E. J., Meerloo, T. & Emr, S. D. Escrt-III: An endosome-associated heterooligomeric protein complex required for mvb sorting. *Developmental Cell* **3**, 271–282 (2002).

70. Adell, M. A. et al. Coordinated binding of Vps4 to ESCRT-III drives membrane neck constriction during MVB vesicle formation. *The Journal of Cell Biology* **205**, 33–49, doi:10.1083/jcb.201310114 (2014).

71. Wollert, T. & Hurley, J. H. Molecular mechanism of multivesicular body biogenesis by ESCRT complexes. *Nature* **464**, 864–869, doi:10.1038/nature08849 (2010).

72. Hanson, P. I., Roth, R., Lin, Y. & Heuser, J. E. Plasma membrane deformation by circular arrays of ESCRT-III protein filaments. *The Journal of Cell Biology* **180**, 389–402, doi:10.1083/jcb.200707031 (2008).

73. Cashikar, A. G. et al. Structure of cellular ESCRT-III spirals and their relationship to HIV budding. *eLife* **3**, doi:10.7554/eLife.02184 (2014).

74. Lata, S. et al. Helical structures of ESCRT-III are disassembled by VPS4. *Science* **321**, 1354–1357, doi:10.1126/science.1161070 (2008).

75. Hierro, A. et al. Structure of the ESCRT-II endosomal trafficking complex. *Nature* **431**, 221–225, doi:10.1038/nature02914 (2004).

76. Thomas, W., Christian, W., Jennifer, L. S. & Hurley, J. H. Membrane scission by the ESCRT-III complex. *Nature* **458**, 172–177 (2009).

77. Angelova, M. I. & Dimitrov, D. S. Liposome Electroformation. *Faraday Discussions of the Chemical Society* **81**, 303–311 (1986).

78. Husain, M., Boudier, T., Paul-Gilloteaux, P., Casuso, I. & Scheuring, S. Software for drift compensation, particle tracking and particle analysis of high-speed atomic force microscopy image series. *Journal of Molecular Recognition: JMR* **25**, 292–298, doi:10.1002/jmr.2187 (2012).

79. Chiaruttini, N. et al. Relaxation of loaded ESCRT-III spiral springs drives membrane deformation. *Cell* **163**, 866–879, doi:10.1016/j.cell.2015.10.017 (2015).

80. Heath, G. R. & Scheuring, S. High-speed AFM height spectroscopy reveals microsdynamics of unlabeled biomolecules. *Nature Communications* **9**, 4983, doi:10.1038/s41467-018-07512-3 (2018).

81. Uchihashi, T., Watanabe, H., Fukuda, S., Shibata, M. & Ando, T. Functional extension of high-speed AFM for wider biological applications. *Ultramicroscopy* **160**, 182–196, doi:10.1016/j.ultramic.2015.10.017 (2016).

82. Ando, T. Directly watching biomolecules in action by high-speed atomic force microscopy. *Biophysical Reviews* **9**, 421–429, doi:10.1007/s12551-017-0281-7 (2017).

6 Reconstituted Membrane Systems for Assaying Membrane Proteins in Controlled Lipid Environments

Gaya P. Yadav and Qiu-Xing Jiang

CONTENTS

6.1 INTRODUCTION

Cell membranes play a pivotal role in separating cells from their environments, maintaining cells in an off-equilibrium steady state, detecting and relaying outside signals into the cells by responding to their environments in specific manners. There has been a strong interest in the scientific community because membrane systems, including both membrane proteins and lipids, are fundamentally important for all live cells, and serve as therapeutic targets for human diseases. Membrane proteins generally include both integral membrane proteins that are integrated in a membrane and traverse both leaflets of the bilayer and peripheral membrane proteins that are associated with lipids or proteins in membranes.[1,2] Around 20%–30% of identified proteins of the human genome are predicted integral membrane proteins.[3–6] Although the total number of integral membrane proteins is increasing over time, determination of the structural basis for their functions still lags behind. In particular, it remains difficult to study quantitatively the effects of membrane lipids on the structure and function of membrane proteins.

Generally, there are several difficulties in the studying a new membrane protein. First, low level of expression or need of over-expression may deter biochemical and biophysical characterization. Almost all membrane proteins, when over-expressed in a heterogenous system, have a lower-than-expected level of expression because of inevitable aggregation[7] or deficiency of proper folding or post-translational modifications in over-expression systems as compared to their native environments. Secondly, difficulty in solubilization of integral membrane proteins may decrease analytical capacity. Since these proteins are integrated inside lipid membranes, suitable detergents must be found to dissolve the membranes and release them into aqueous phase. Different cellular or organellar membranes have different lipid compositions such that a specific type of detergents may be needed to solubilize the membrane and release the proteins of interest. A comprehensive review on the importance of detergents in membrane protein solubilization and detergent-protein interactions was presented before by le Maire et al.[8] It also provided an overview of techniques that could be employed for the structural investigation of detergent-solubilized membrane proteins.[8,9] Despite these difficulties associated with handling membrane proteins, their studies remain a fundamental area due to their roles in different biological processes and their serving important therapeutic targets for various human diseases. After purification of an integral membrane protein in detergents, it is usually necessary to test its functionality, which usually can only be performed well in membrane, not in detergents. Reconstituted membranes with the target proteins, even though quasi-physiological in mimicking native membranes, are important tools for studying membrane proteins in well-controlled lipid environments. In this chapter, we will introduce methods for reconstitution of membrane proteins and multiple functional assays commonly used in our laboratory and others for quantitative evaluation of functionality of membrane proteins in reconstituted membrane systems. At the end we will present personal perspectives on what limitations current techniques harbor and what solutions may become possible in the future.

6.2 RECONSTITUTION OF MEMBRANE PROTEINS INTO MODEL MEMBRANES

6.2.1 PREPARATION OF PROTEOLIPOSOMES

Proteoliposomes may contain enriched membrane proteins in either native or artificial membranes. Microsomes prepared from ER membranes have been extensively used for studying *in vitro* translation and co-translational translocation of proteins (e.g., [10,11]). In intact cells, flux of fluorophores through native biomembranes or isolated organelles, such as mitochondria or microsomes, could be directly employed to study the functionality of membranes proteins, especially transporters.[12–14] The most commonly applied systems are cells expressing endogenous transporters or immortalized cells from the same tissues that over-express target membranes proteins. This method has advantages of containing native proteins in cell membranes and possibly their interacting partners which might be important for their function. In the latter case, the secondary cultures may still have lipid environments that are different from the primary cells from the original tissues. On the other hand, transient or stable over-expression systems such as *X. leavis* Oocytes or tumor cell lines have the advantage of producing a large quantity of the proteins of interest. The intact expression system allows identification and characterization of a large number of membrane systems from plasma and/or intracellular membranes. However, the complexity of the secondary expression systems is associated with the interference from homologous or contaminant proteins, which could give rise to false positive results. The application of specific inhibitors is important in these systems to separate the activities of the target molecules. But for intracellular proteins in intact membrane systems, inaccessibility of inhibitor-binding sites may limit the efficacy of specific inhibitors in reaching the targeted membrane proteins. Further, incomplete separation of different intracellular membranes and plasma membranes in cellular fractionation usually makes it difficult to achieve a high level of certainty in localizing the protein function to the right compartments. Genetic manipulation and high-resolution electron microscopy are usually required. In general, the above-mentioned limitations have introduced uncertainty or sometimes caused controversies into varying aspects of investigating the structure, function and regulation of specific membrane proteins. For example, ClC-3 Cl^-/H^+ exchanger was proposed by some to be present in insulin-secretory granules,[15] which was refuted by others based on antibody specificity.[16]

To overcome the complications and limitations of native microsomes and intact cell membranes, lipid vesicles prepared with purified integral membrane proteins in bilayers of well-defined lipid composition are an excellent tool for mechanistic studies of membrane proteins.[17] These vesicles are called proteoliposomes. Proteoliposomes should satisfy several criteria before being considered useful for such purposes.[18,19] The usual criteria are the following.

1. Proteoliposomes should be of relatively uniform size distribution which can be achieved by extrusion through membranes of a selected pore size. Such an operation usually produces vesicles with a quasi-Gaussian distribution in dimeter. Roughly uniform dimensions avoid the activities in vesicles larger in volumes or surface areas dominate the assays to be conducted.

2. The protein should be evenly distributed among the proteoliposomes. With thorough mixing of proteins and lipids before detergent removal, this condition is usually satisfied. The population of vesicles with a specific number of protein molecules follows a Poisson distribution.

3. The proteins must be reconstituted at different lipid-protein molar ratios (PLR) so that the protein activity as a function of average number of protein molecules per vesicle may be quantified in order to deduce possible cooperativity among the protein subunits.

4. The biological activity must be high in the reconstituted proteoliposomes. With well reconstituted proteins, this is a problem in signal-to-noise ratio (SNR). Robust assays must have high SNR and be conclusive about the functionality of the reconstituted proteins.

5. Reconstituted liposomes must be tightly packed with no residual detergents such that the ionic compositions across the vesicle membranes could be varied to generate ionic gradients. A Ca^{2+}-packaging experiment may be performed as a positive control in testing the success of loading Ca^{2+} into vesicles and thus the tightness of the vesicular membranes. Ca^{2+} release from vesicles can be monitored with high-sensitivity Ca^{2+} dyes, such as calcium green 1, Fura-2, etc.

6. For proteins with unidirectional transport of ions,[20] vectorial insertion or preferred orientation of the membrane protein in liposomes will be required for robust signal. Orientational preference in vesicles happens sometimes, but almost always incompletely. When the targeted membrane proteins have a geometric shape (conical) to favor one orientation over the other, or when the ectodomains in one side of the transmembrane domain is significantly larger than that of the other, larger ectodomains are preferentially positioned outside of the vesicles for thermodynamic reasons. For more complete orientation selection, bead-supported unilamellar membranes (bSUMs) are suitable for guiding preferential insertion and will be discussed later.[21]

There are multiple ways to prepare proteoliposomes.[22] The first is a dilution method in which the protein-detergent mixture is diluted such that the liposomes will form when detergent concentration decreases below its critical micellar concentration (CMC). The proteins will be inserted into liposomes containing detergents. Further removal of the detergents from the vesicles will produce well-packed proteoliposomes.

The second method is to produce a thin layer of lipids on the inner surface of a round flask. After the lipids are dried, membrane proteins in detergents with a detergent concentration slightly above its CMC are added. With gradual dissolution of dried lipids, the proteins are spread into liposomes with a specific ratio of protein: detergent: lipid. The detergents can then be removed by BioBeads, gel filtration chromatography or dialysis.

The third method is to dislodge dried lipids into the suspension, sonicate the suspension and then completely solubilize the lipids with detergents before the preparation of protein: lipid: detergent mixture in a specific ratio. Removal of detergents from the mixture can be done the same.

Complete detergent removal is critical for producing tightly packed vesicles. We will describe a more generic method which starts with a mixture of protein/detergent/lipid. A mixture of PE/PG (POPE/POPG) will be used to mimic bacterial cell membranes,[17] and a mixture of PC: sphingomyelin: cholesterol can be utilized to mimic a eukaryotic membrane if needed.[23] A step-by-step operation follows:[17]

1. Transfer 3.75 mg of POPE and 1.25 mg of POPG (POPE: POPG = 3:1 in weight ratio) in chloroform into the pre-cleaned, screw-capped glass test tube and dry the lipids under a continuous stream of argon gas. Rotate the testtube during this process so that the lipids are more or less evenly distributed on the inner surface of the tube. Afterwards, keep the test tube under vacuum for an hour to ensure maximal removal of chloroform.

2. Add an adequate amount (a final lipid concentration 10.0 mg/mL with protein and other constituents) of low salt-buffer (10 mM HEPES, pH 7.4) or water into the dried lipid mixture, and vortex the test tube to hydrate and dislodge the lipids. The lipid suspension looks whitish and turbid.

3. Add some argon gas on top of the rehydrated lipids to minimize oxidation of lipids.

4. Sonicate (**Note #3**) the lipid suspension in an iced water-bath sonicator until the vesicle solution becomes translucent (OD_{410} < 0.2). Typically for a 10 mg/mL PE/PG solution in water or low salt buffer, sonication of 30 seconds ON and 30 seconds OFF (on ice) for a total of 15 minutes will be enough. Longer sonication is not recommended due to possible damage of lipid molecules. The POPE/POPG vesicles should become nearly transparent and look azure against white light. The absorbance at 410 nm (OD_{410}) can be measured at this stage to examine whether the vesicles are small in diameter and thus low in light-scattering. For PE/PG vesicles, OD_{410} may reach <0.1 after sonication. OD_{410} varies with lipid composition. For certain lipid mixture it may be difficult to reach low OD. If that happens, it is recommended to use detergents and allow long incubation to reach nearly complete solubilization of lipids and good distribution of lipids around the protein-detergent micelles.

   Alternatively, a microprobe sonicator in contact with the lipid suspension in a plastic tube can be introduced to produce the same results. To avoid lipid oxidation on the surface of the microprobe, it is important to keep the sample cooled, and add some reducing agents (1.0 or 2.0 mM DTT or TCEP). When the lipid solution is nearly translucent, the size of the SUVs is in the range of 30–50 nm. The curvature of the SUVs is so high that a small perturbation (e.g., with a small amount of detergents) is enough to induce vibrant fusion of these vesicles into larger ones that are 80–200 nm in diameter.

5. Add KCl or NaCl, based on the proteins to be reconstituted, to a final concentration of 300 mM and 20–25 mM of DM or other detergents in which the proteins are purified. Incubate the solution with horizontal rotation for 1.0–2.0 hours at room temperature to form evenly distributed lipid/detergent mixed micelles.

6. Add target proteins at a protein: lipid molar ratio of 1:10,000 or another ratio, depending on the stoichiometry of the functional protein unit, to incorporate certain numbers of functional units into individual vesicles. A range of PLR of 1:25, 1:50, 1:100,1:500, 1:1,000, 1:10,000 (see **Note 4**) may be tested to study the possible cooperativity among the subunits within each functional complex. For the DM/PE/PG mixture, 10 mM HEPES (pH 7.4), 300 mM KCl, NaCl or other required salts and detergents to final concentration of 25 mM (DM) or a different detergent at a concentration required to solubilize the lipid vesicles completely into lipid-detergent micelles. To characterize the solubilization itself, OD_{410} can be used to follow vesicle solubilization as a function of detergent concentration. The resulted protein/detergent/lipid mixture is incubated at room temperature or cold room for 2.0–3.0 hours with end-over-end rotation to reach near-equilibrium distribution of the three components. The selection of detergent concentration is guided by the detergent-induced vesicle fusion and vesicle solubilisation.[24] For the reconstitution of a newly identified membrane protein, it is advised to start with the PC extracts, such as *E. coli* polar lipid extract, soybean PC extract, egg PC extract, etc., such that if certain phospholipids are needed for its function, they may be already included in the lipid mixture.

7. Removal of detergents from protein/lipid/detergent mixture reconstitutes the protein into vesicles. In order to achieve better reconstitution efficiency, it is important allow enough time for proteins and lipids to interact while the detergent concentration falls slightly below its CMC. We therefore usually avoid using gel filtration chromatography to remove detergents. Instead, either dialysis or treatment with hydrophobic BioBeads is preferred. We will describe the two in the following.

   a. Dialysis for gradual removal of detergents. A piece of dialysis tubing (e.g., Spectrum labs) with a pore size suitable to retain protein molecules above 10 kDa is chosen. It separates detergent monomers from the protein/lipid/detergent mixture due to slow diffusion of the former across the tubing. It is usually a good idea to keep the molecular weight cutoff smaller than the micelle size so that the lipid/detergent mixed micelles will not be lost through diffusion. The tubing is washed with a lot of DI water and is boiled in a low-salt buffer containing 2.0 mM EDTA (10 mM Tris-HCl/ HEPES of the required pH) for ~20 minutes. Sterilization and removal of divalent cations (mainly Mg^{2+}) are introduced to protect the target proteins. After cooled down, the tubing is washed with the dialysis buffer, which is the same as the solution to be loaded inside the vesicles. The protein/lipid/detergent mixture is then loaded into the dialysis tubing with minimal air space left. Afterwards, the assembled tubing is floated inside the dialysis buffer with continuous stirring. The volume ratio between the protein/lipid/detergent mixture and the dialysis buffer is usually kept at ~1:1000. The buffer is changed every 4–8 hours and the vesicles are ready after five times of buffer change. During the dialysis, it is recommended to rub the dialysis tubing every couple of hours and rotate the tubing end-over-end to avoid clogging of the pores. The proteoliposome suspension becomes turbid because the diameter of the formed vesicles ranges from 80 to 300 nm, making them good scatters of visible light. Sometimes, there might be obvious whitish precipitates falling to the bottom inside the dialysis tubing, which usually are multilamellar vesicles (MLVs). Rubbing the tubing during dialysis may help decrease these precipitates. Keeping a detergent concentration of 0.5 CMC during the first round of dialysis may help minimize MLVs. After dialysis, the proteoliposomes are collected, aliquoted, flash-frozen in liquid nitrogen and stored at −80°C until their use.

b. BioBeads with hydrophobic pores are useful for removal of detergents with a very low CMC[25–34] and thus sometimes can be added to the vesicles formed by dialysis to remove residual detergents. The hydrophobic pores allow the hydrophobic tails of detergents to be trapped inside for removal. Before suitable for detergent removal, the BioBeads are first cleaned with methanol and sonicated briefly, and then washed with MQ water for 3–4 times before being stored in 20% ethanol. Due to microporous nature of the BioBeads, air sometimes gets trapped within the pores, making some of the beads floating. Methanol wash plus sonication helps remove trapped air, an important step for making the BioBeads usable. Degassing the beads under vacuum may also be applied to remove trapped air.

There are two ways to use the BioBeads—column method and batch method. For the column method, 60%–70% slurry of BioBeads are loaded into a column. After drainage of water and 2–3 times of wash with the required buffer, the sample (protein/detergent/lipid mixture) is loaded, and the column is washed with the reconstitution buffer in a slow flow rate (0.2–0.3 mL/minutes). The column can be regenerated by washing with water, 100% methanol, water and buffer. For the batch method, the BioBeads stored in 20% ethanol are washed with water and the targeted buffer. After removal of excess buffer, the wet beads are weighed out and mixed directly with the protein/lipid/detergent mixture. For every batch treatment, the wet eight of the beads is 20–60 times of the total detergent weight in the mixture, and the incubation time is 15–30 minutes at room temperature or 4.0°C. To change beads, the protein mixture is moved to a fresh tube with fresh beads. Based on the estimated binding capacity of the beads for a particular type of detergent, the total weight of beads to be used is ~100-folds of the total detergent weight,[35,36] and can be divided into five batches of 10%, 20%, 20%, 20%, and 30% of the total. Each batch of beads needs 20-minute incubation time at room temperature or 30–40 minutes in a cold room. The last step can be longer to remove the trace amount of detergents. To avoid protein degradation, protease inhibitors can be introduced when they do not interfere with the bead treatment. The proteoliposomes should be collected by pipetting carefully without taking the absorbent beads, aliquoted, flash-frozen, and stored at −80°C until use. The batch method has been used by most of the researchers for the sake of convenience.

The reconstitution methods described above are widely used and effective in most cases. Proper integration of membrane proteins into vesicles can be tested by floating them from the bottom of a density gradient to the top. A three-step gradient made of 10%, 35%, and 55% of sucrose or 5%, 10%, and 15% Ficoll 400 in the reconstitution buffer has worked well.[17,37] It is well known that this method cannot control the final orientation of membrane proteins in proteoliposomes.[38]

6.2.2 RECONSTITUTION OF MEMBRANE PROTEINS INTO NANODISCS

Small proteoliposomes (say less than 100 nm in diameter) have intrinsic curvature which may affect the protein function. Nanodiscs offer a solution to insert proteins into a small patch of planar membrane (**Figure 6.1**). A nanodisc is composed of a patch of phospholipid bilayer with its hydrophobic edge shielded by two copied of an amphipathic scaffold protein. The nanodiscs are a particularly attractive option for studying membrane proteins in the context of ligand-receptor interactions and are very useful for controlling both sides of the transmembrane proteins in a bilayer. Sligar's group developed the method for the reconstitution of membrane proteins into nanodiscs.[39–41] They took advantage of the special property of naturally occurring amphipathic apolipoprotein A-1 of high-density lipoproteins. Detergent-solubilized phospholipids,

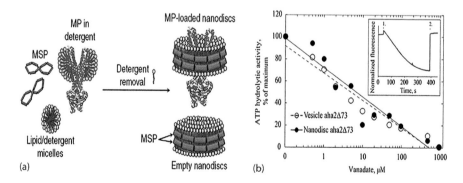

FIGURE 6.1 Reconstitution of membrane proteins in nanodiscs. (a) Schematic presentation of the assembly of empty and target nanodiscs from a mixture of MSP, lipid/detergent mixed micelles and detergent-solubilized membrane proteins (MP) upon detergent removal. (Adapted from Zoghbi et al., Nanotechol Rev. 6(1), 33–46, 2017. With permission.) (b) ATP hydrolytic activity of P-type H^+-ATPase (aha2Δ73) reconstituted in vesicles (open circles) and Nanodiscs (filled circles) under varying concentration of vanadate. Data are presented as fractions of the control activity measured in the absence of vanadate. IC50 values for vanadate measured under two conditions, 5.8 μM for proteins in vesicles and 4.4 μM for enzymes in nanodiscs, indicate that proteins reconstituted in nanodiscs are active as those in vesicles. Inset, confirmation of proton pumping of aha2Δ73 in vesicles using a proton-dependent fluorophore ACMA. Pumping was initiated by adding $MgSO_4$ (peak 1) and the resulted proton gradient was dissipated by addition of CCCP. (Adapted from Justesen, B.H et.al., *J. Biol. Chem.*, 288, 26419–26429, 2013. With permission.)[46]

membrane scaffold protein (MSP) and proteins of interest are mixed in an empirical ratio. After incubation for 1–2 hours, detergents are removed using BioBeads or dialysis. The MSPs shield the edge of the bilayer formed around the transmembrane protein. The nanodiscs may vary in diameter from 9.8 to 17 nm, depending on the MSPs (**Figure 6.1**), and can accommodate different sizes of transmembrane proteins.[39,41–45] The nanodiscs have advantage over reconstitution vesicles by keeping the proteins in a more native-like environment. Nanodiscs are widely used for the structural studies using cryoEM. There are more than 100 structures of membrane proteins in nanodiscs. A detailed procedure for the preparation of nanodiscs has been described by Sligar's Lab.[42,43] We have adapted the Sligar's procedure with the detergent removal methods described in the last section and have been successful in making nanodiscs.

A note of caution here is that the nanodiscs may allow the interaction of the MSPs with the membrane proteins such that the proteins may behavior differently from those in a large continuous membrane. For certain membrane proteins, the amount of lipids around their transmembrane domains might be very small, which might beat the purposes of the reconstitution. Further, the transmembrane movement of ions or other moieties through or along the transmembrane proteins is difficult to measure in nanodiscs.

6.2.3 RECONSTITUTION OF TRANSMEMBRANE PROTEINS INTO BEAD-SUPPORTED UNILAMELLAR MEMBRANES (bSUMs)

A supported bilayer is another way of studying biochemical and biophysical properties of membrane proteins. In the past, supported bilayers on planar slides both as a single bilayer or multiple stacked bilayers have been prepared.[47–51] This method was first introduced by Tam and McConnell in 1985. A supported planar bilayer is attached to the solid support on one side, and the other side is exposed to the aqueous solution. The attachment of the lipid headgroups to the solid surface, however, may significantly limit lipid mobility in the attached leaflet, and in some cases may affect the biological activity of the reconstituted membrane proteins, especially for those having large ectodomains on both sides of their transmembrane domains. Some of these problems may be partially alleviated by bilayers tethered onto polymer cushions.[6,52] Integration of small proteins has been successful using polymer cushions with measurable biological activity, but the polymer-tethered bilayers are often not suitable for large membrane proteins. To overcome these problems, Zheng, H. et al.[21] developed a new method, which uses membrane proteins tethered on a bead surface through chemical linkage to support the incorporation of different lipids that may otherwise be difficult to introduce into bilayers (Figure 6.2). For example, it is usually difficult to make good planar bilayers composed of a mixture of PC: sphingomyelin: cholesterol (say 5:1:1 in molar ratio) using the decane solvent system. Instead, the tethered membrane proteins can guide the formation of a bilayer membrane around each bead. They named this reconstituted system a bead-supported unilamellar membrane (bSUM).[21] The bSUM has two major advantages: (1) directional (vectorial) insertion of membrane proteins in bilayers through chemical linkage, and (2) precise control of lipid composition. For example, Zheng et al. were successful in introducing cholesterol into the bSUMs, which are suitable for electrical recordings of channel activity using planar glass electrodes. This system

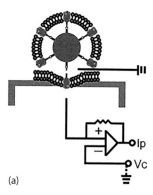
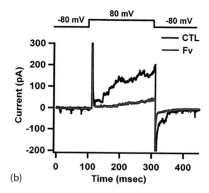

(a)   (b)

FIGURE 6.2 Chemical principles behind the formation of bSUMs. (a) After surface functionalization through chemical oxidization and bioconjugation, the beads were incubated with membrane proteins containing active ligands, such as His tags, and detergent-solubilized lipids. The molar ratio of protein: lipid: MSP was tested and for KvAP it was 1:100:2. The detergent from the mixture was removed with BioBeads (SM-2, BioRad). When the vesicles of bSUMs are attached to a planar glass electrode, the vesicle membrane ruptures by making contact with the glass surface, leaving the top partition opened to the intravesicular side of the bSUM, which is similar to an excised patch (inside-out mode). (b) Electric recordings of the tetrameric KvAP channels in bSUMs. Inhibition of the channel activity by a highly specific immunoglobular domain, Fv. As a result of membrane disruption on the top partition, recordings were made in the inside-out mode (black trace; the bottom partition is grounded). When 100 µg/mL Fv was perfused into the extravesicular side of the channels (facing the bottom partition) for 2.0 minutes at 0 mV, the channel activity was almost completely blocked (grey trace). (Adapted from Zheng et.al. With permission. http://jgp.rupress.org/content/147/1/77.)

allowed them to quantitatively characterize the effects of cholesterol on a voltage-gated potassium (Kv) channel in a well-controlled lipid environment.[35,36]

6.3 ACTIVITY ASSAYS FOR RECONSTITUTED MEMBRANE PROTEINS

Once targeted membrane proteins are successfully reconstituted into proteoliposomes, nanodiscs or bSUMs, the next question is whether the proteins in these systems retain their biological activities, and if so how to measure them in a quantitative fashion. Here, we will discuss different methods developed for this purpose. We will not discuss functional analysis of ion channels by patch clamp recordings from cell membranes or black lipid membranes, or by measuring ion flux through channels in cells. Instead, we will discuss the methods we and others have developed or implemented for reconstituted proteoliposomes.

6.3.1 Ion Flux Assays

Ion channels conduct ions across membranes with a high flow rate. If they have high preferences for a specific type of ion, they are named after that ion. Some of them conduct cations and are called cation channels while others are called anion channels. Still others may conduct neutral molecules, such as water, ammonium, glycerol, urea, etc. At any specific moment, the flux (flx) of an ion across an ion channel can be described by

$$flx = g*(Vm - Vrev),$$

with g representing the single channel conductance of the specified ion, Vm the transmembrane electrostatic potential and $Vrev$ the reversal potential for the ion. Ion flux is a direct measurement of the activity of ion channels. Because direct electric recordings usually take more time than biochemical assays, ion flux assays are preferred, although less informative than detailed single channel analysis. Radiolabeled ions were used for measuring flux before. But recent advancement made it possible for optical measurement to be used to monitor ion flux through ion channels. Next, we will describe two specific assays we have implemented in the lab.

6.3.2   LIGHT-SCATTERING-BASED ION FLUX ASSAY

This method was first introduced by Jin, A. J. for extruded liposomes[53] and later was used for studying CLC antiporters[54] and for the chromogranin B (CHGB) chloride channel that normally functions in regulated secretion.[25] After reconstitution of the proteins, the proteoliposomes are extruded through a porous membrane filter with 400-nm holes in order to make them homogenous in the presence of a specific, high salt concentration. Right before use, the extravesicular ions are replaced by desalting size-exclusion chromatography in order to create ionic gradients across the vesicular membranes. Due to the ionic gradients between intra-vesicular (high) and extravesicular (low) solutions, the ions flow out of the vesicles and cause deformation of the vesicles, which will cause a significant increase of light scattering (Figure 6.3). The change in deflected light can be used to estimate indirectly the ion flux across the reconstituted ion channels. The following is a typical step-by-step procedure using CHGB anion channels as an example.

1. Prepare CHGB proteoliposomes as described above with 300–500 mM of specific ions inside. For example, KCl is used in the diagram of Figure 6.3.
2. Wash a manual extruder with MilliQ water followed by an extravesicular buffer containing 10 mM HEPES-KOH, pH 7.4, 300 mM K-isethionate.
3. Soak a porous membrane with 400 nm pores and two support meshes in the extra-vesicular buffer for 10 minutes (Avanti polar lipids). Rinse the extruder with the extra-vesicular buffer. Assemble the extruder with the filter membrane and two support meshes on both sides of the membrane to avoid membrane breakage.
4. Rinse syringes with the buffer and extrude the extra-vesicular buffer a few rounds across the membrane to remove trapped air bubbles between the membrane and the support meshes. Discard the buffer and fill one injection syringe with 1.5 times of the final amount of proteoliposomes

Schematic diagram depicting flux assay

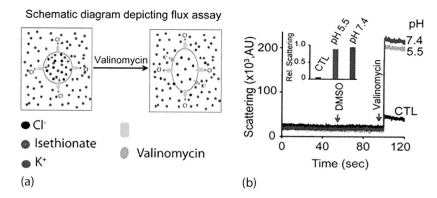

● Cl⁻
● Isethionate
● K⁺   ● Valinomycin

(a)   (b)

FIGURE 6.3 A light-scattering-based anion flux assay. (a) Purified CHGB was reconstituted into lipid vesicles in the presence of high KCl. Extravesicular Cl⁻ was replaced with isethionate, leading to a strong Cl⁻ gradient across the vesicular membrane. After a small leak through the Cl⁻ channel, a positive potential developed inside, and the leak was stopped. Addition of valinomycin, a K⁺ ionophore, dissipated the positive potential and allowed continued efflux of Cl⁻. The sudden efflux of KCl caused the shape change of vesicles and increased light-scattering at 600 nm (Ex: 600 and Em: 600 nm). Left: an extruded and salt-exchanged vesicle was changed into a buffer of K-isethionate. Right: Adding valinomycin causes a shape change and increases average Stokes radii of the vesicles. (b) Representative traces of light scattering changes in an ion flux assay. Red and cyan traces show that Cl⁻ efflux leads to increased scattering when CHGB channels are open in both pH 7.4 and 5.5. The black trace shows the vesicles without CHGB channels have no activity. The inset bar graph depicts relative change in scattering under different conditions. (Adapted from Yadav, G.P. et al., *Life Sci. Alliance*, 1, e201800139, 2018. With permission.)

required for planned experiments. It is common to lose some of the liposomes during extrusion. With a small amount (~100–200 μL) of vesicles, it is advised to cut the 1.5 mL microcentrifuge tube from top so that the tip of the extruder syringe can reach the bottom in order to avoid air bubbles.

5. Extrude proteoliposomes for 21 rounds with a gentle pressure to avoid heating and breakage of the porous membrane. For the initial couple of rounds, a little more pressure may be needed. Pressure would decrease during later rounds because the majority of vesicles are extruded into a size slightly smaller than the average pore size. Collect extruded vesicles opposite to the loading side to avoid giant vesicles which were not extruded well.

6. Load extruded vesicles to a pre-equilibrated desalting column (Sigma or Bio-Rad). One can pack a size exclusion column using pre-soaked G-50 resin (**Note #1**). Collect the vesicle fractions from the size-exclusion column and immediately use them for flux experiments (**Note #2**).

7. Dilute the buffer-exchanged vesicles in a degassed extra-vesicular buffer and measure 90° light scattering at 600 nm inside a fluorometer (excitation and emission both at 600 nm).

8. Adding an ionophore (1.0 μM; for example, valinomycin for K^+ or nigericin for H^+; see **Note #2**) will allow ion efflux to happen. The sudden efflux of ions makes the initially round vesicles deform, leading to a substantial increase in light scattering. Change in the scattered light is measured, representing the channel activity[25,54] (Figure 6.2).

6.3.3 FLUX ASSAY USING A STOPPED-FLOW SYSTEM

The above bulk-phase assay is a conventional steady-state assay because it only measures steady-state activity due to poor time resolution. For a more quantitative measurement of the early phase of the ion flux, a stopped-flow system can be employed to avoid the time lapse between the addition of the ionophores and the measurement of light scattering. Next, we will describe a procedure for measuring ion flux in a stopped-flow system.[25]

1. Prepare the SX20 (or similar) stopped flow fluorometer system by initializing the system, light sources and temperature controller and setting excitation and emission wavelengths to 610 nm. Please make sure that a filter of a proper wavelength is used.

2. Wash the fluid handling system with MQ water followed by the buffer to be used for activity measurement of the protein of interest (three times each) carefully to avoid any visible bubble in the syringe and in the tubing. Set the software to record 1 second of data with 5000 recorded points (0.2 ms each point) and set the delay time between sequential sample mixing events to a desired value.

3. Right before each experiment, exchange vesicles with a buffer containing desired membrane-impermeable anions (such as isethionate), and then load them to the prewashed syringe. The other syringe contains valinomycin in the reaction buffer (e.g., isethionate buffer). It is very important to remove all bubbles in the syringes during loading and to degas the reaction buffer (e.g., 10 mM HEPES pH 7.4, 300 mM K-isethionate).

4. Dilute the vesicles to a concentration that is two times of the final reaction concentration. The suitable concentration for reaction should be tested beforehand so that a robust signal can be obtained.

5. Prepare 2.0 μM valinomycin in the assay buffer with a minimum concentration of DMSO and load it into one of the injection syringes.

6. After fast mixing (with a dead time of say 2.0 ms), light scattering at 610 nm is monitored. Signals from three to five consecutive scans should be averaged to gain sufficient statistical power. Repeating the experiment multiple times makes the signals highly reproducible. The results can be represented as a percentage change in the scattered light with and without the proteins in the liposomes. It is advisable to use another membrane protein as negative control for the experiment.

7. Repeat **steps 4–6** for all experimental conditions of interest. For example, perform titrations of activating ligands and/or channel blockers in the reaction buffers. Alternatively, use a specific set of conditions and vary the delay

time between consecutive series of mixing reactions to determine how the channel activity changes as a function of mixing time with the reaction buffer.

8. Examine the data and discard traces with obvious mixing artifacts. Normalize the light-scattering data against that from the control sample and use the flux rates from all experimental conditions to analyze ligand dose-response curves and/or ligand-binding kinetics.

6.3.4 MEASURING CHLORIDE OR FLUORIDE EFFLUX BY AN AG/AGCL OR AG/AGF ELECTRODE

For anions, it is feasible to measure anions using highly specific electrodes that are sensitive to changes in their concentrations. Ag/AgCl and Ag/AgF can be used to measure Cl⁻ and F⁻, respectively.[54] In next we will describe the measurement of Cl⁻ release from CHGB-vesicles.[25]

1. Prepare the extruded proteoliposomes as described earlier. After substitution of extravesicular Cl⁻ with isethionate, proteoliposomes are diluted by 20 x into an extra-vesicular buffer containing a small amount (0.20 mM) of Cl⁻.

2. The ground Ag/AgCl is connected to the recording chamber through a salt bridge. The recording Ag/AgCl electrode is directly immersed into the recording solution. ~0.2 mM KCl is added to the recording solution to stabilize the baseline (**Figure 6.4**).

3. DMSO is added as control before valinomycin in DMSO is introduced at 0.25–1.0 µM to trigger the efflux of Cl⁻. A stirring bar is used to mix valinomycin with the vesicles well.

4. Currents are recorded in the whole-cell mode using a patch clamp recording system (Axon 200B) with $V_m = 0$ mV and the lowest possible gain. The analog data are filtered at 200 Hz and sampled at 2 kHz. Gap-free recordings are made for 45–60 seconds. At the end of the experiments, 50 mM β-octylglucoside (β-OG) is added to release all chloride ions. Vesicles without protein or with a control protein should be used as negative control (top panel in Figure 6.4b). Recorded signals can be normalized against that after detergent addition and the signal from control protein vesicles or empty vesicles without channel protein can be subtracted after normalization (Figure 6.4).

6.3.5 FLUORESCENCE-BASED FLUX ASSAY

Another way to measure ion fluxes is to use fluorescent dyes that are sensitive to specific ions or to change in transmembrane potential. The success of these assays may make them suitable for high-throughput screens against specific channel openers or inhibitors. We will describe them in the next by using examples from literature.

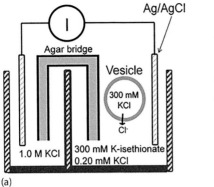
(a)

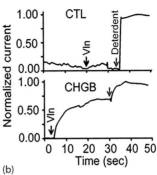
(b)

FIGURE 6.4 Direct measurement of chloride release from reconstituted vesicles. (a) Schematic diagram showing the experimental setup for the direct current measurement using Ag/AgCl electrode. CHGB vesicles are loaded with 300–450 mM KCl. To begin the assay, extravesicular Cl^- is removed and vesicles are diluted into low Cl^- solution (~0.2 mM), which creates a high gradient for the efflux of Cl^- ions. Continued chloride efflux is initiated by 0.5 μM valinomycin (Vln). Change in the current is recorded in the whole-cell mode with $Vm = 0$ mM and at the lowest gain. (b) A typical trace of real experiment using the Ag/AgCl electrode to measure Cl^- release from the CHGB reconstituted proteoliposomes. Cl^- release from vesicles was recorded with Ag/AgCl electrode. At the end 10 mM β-OG was added to release all the Cl^- ions. Top: recording from control vesicles without CHGB; Bottom: recoding from CHGB vesicles. (Adapted from the Yadav, G.P. et al., *Life Sci. Alliance*, 1, e201800139, 2018. With permission.)

6.3.5.1 Fluorescence Assay Using ACMA for Potassium Channels

ACMA (9-amino-6-chloro-2-methoxyacridine) is a DNA intercalator that selectively binds to poly (d(A-T)).[55] Unprotonated ACMA can permeate across membranes. At an acidic pH, it is protonated and becomes membrane impermeable. Protonation quenches AMCA fluorescence.[56] The movement of cations and anions across the membrane may be coupled with the movement of protons, which generates a proton gradient and can be monitored with ACMA.[56] ACMA has been used to study the proton-pumping activity of various membrane-bound ATPases before.[57]

We will use K^+-channel as an example to explain the use of ACMA. The principles behind the experiment are shown in **Figure 6.5**. With K channels in vesicles, K^+ efflux is coupled to proton influx into the vesicles when there is a proton conductance. CCCP is a proton ionophore and can be added into the vesicles to allow proton influx to happen. Upon acidification of the vesicle lumen, ACMA is protonated inside vesicles and becomes enriched and trapped inside. At the same time, ACMA fluorescence inside vesicles is quenched, leading to a significant decrease of total ACMA fluorescence. The coupled movement of K^+ (efflux) and H^+ (influx) can thus be followed by the quenching of ACMA fluorescence. Here the rate-limiting state in fluorescence change is the movement of ACMA across the membrane. A detailed procedure for the assay will be presented in the following.

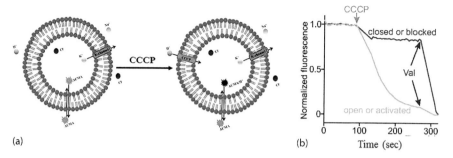

(a)   (b)

FIGURE 6.5 ACMA-based flux Assay. (a) A diagram showing the basic idea. Purified K+
channels are reconstituted in lipid vesicles loaded with high KCl. To start the assay, vesicles
are diluted into a high NaCl solution, creating a strong gradient to drive K+ efflux. Addition of
the proton ionophore carbonyl cyanide m-chlorophenylhydrazone (CCCP) can counterbalance
the negative potential caused by K+ efflux. The H+ influx can be monitored by H+-dependent
quenching of the fluorescence of ACMA. Neutral ACMA is permeable to membrane but
becomes nonpermeable once protonated. Protonation quenches ACMA fluorescence significantly. Acidification of the interior of the vesicles traps protonated ACMA inside. (b)
Representative traces for a typical assay. Green trace shows that K+ channels are open while
red trace shows the closed or inhibited channel. Valinomycin (Val) is added last to allow
complete release of potassium ions. (Panel B was adapted from Su, Z. et al., *Proc. Natl. Acad.
Sci. USA*, 113, 5748–5753, 2016. With permission.)

1. Reconstitute channel proteins into liposomes as described before by using
   POPE (1-palmitoyl-2-oleoyl-sn-glycero-3-phosphoethanolamine) and POPG
   (1-palmitoyl-2-oleoyl-sn-glyvero-3-phopo-1′-rac-glycerol) lipids in a 3:1
   weigh ratio. Dialysis is used for complete detergent removal. Keep the salt
   (K+ ion) concentration high (~400 mM) during reconstitution. Please note
   that for certain lipids (such as cholesterol), BioBeads may change lipid composition of the vesicles (unpublished data) due to bead binding. In such cases,
   dialysis is a better option.
2. Dilute proteoliposomes (by 100–200 fold) into an extravesicular buffer (low
   K+ buffer; e.g., 150 mM NaCl). All other constituents remain the same as
   in the dialysis buffer. For example, if vesicles are prepared at 10 mg/mL of
   lipid concentration, adding 10 μL of the vesicles to 1.0 mL of extravesicular
   buffer (10 mM HEPES pH 7.4, 150 mM NaCl) will initiate the efflux of a
   small number of K+ ions through the reconstituted K channel, which would
   be stopped quickly due to the negative potential inside.
3. Add 1.0 μM ACMA in the reaction mixture and start measuring its fluorescence inside a fluorometer with an excitation wavelength at 410 nm and
   an emission wavelength at 485 nm. The slit size could be adjusted according to the amplitude of the signals. Once the baseline is stable, indicat-
   ing a stable distribution of ACMA across membranes, add 2.0 μM CCCP
   to counter-balance charges accumulated due to K+ efflux. Continued K+
   efflux allows a continued H+ influx and thus acidification of the vesicle
   interior.

4. H$^+$ influx into the liposomes quenches ACMA fluorescence, which will shift the balance for more neutral ACMA to move into the vesicles and become quenched and trapped inside. The change in fluorescence is recorded to reflect the average activity of K$^+$ channels in vesicles. Please note the signal is an indirect measurement of ion movement and much slower than the ion movement across membranes.

5. Once the fluorescence signal is stable, add 1.0 μM valinomycin to maximize K$^+$ efflux, leading to the maximal signal from ACMA quenching. The signal after valinomycin addition is used as an internal control and for normalization of the K-channel-mediated K$^+$ efflux.

The ACMA-based flux assay can be adapted for high-throuput drug screening to discover channel-specific inhibitors or activators.[58]

6.3.5.2 Liposomal Fluorescence Assay Using Lucigenin for Cl$^-$ Ion Channels

Lucigenin is a low-sensitivity fluorescent indicator for Cl$^-$.[59] It can be quenched via collision with Cl$^-$ and has been successfully used for glycine receptors and CFTR channels.[60,61] It is thus suitable for monitoring the change of Cl$^-$ concentration inside proteoliposomes or in the extravesicular medium. We will describe a case for the latter with lucigenin in the outside (Figure 6.5).

1. Reconstitute purified proteins into proteoliposomes using POPE (1-palmitoyl-2-oleoyl-sn-glycero-3-phosphoethanolamine) and POPG (1-palmitoyl-2-oleoyl-sn-glycero-3-phopo-1′-rac-glycerol) lipids in 3:1 weigh ratio and dialysis to remove detergents. Keep the salt (Cl$^-$ ion) concentration high (~400 mM) during reconstitution.

2. Immediately before the flux assay, remove the extravesicular chloride ions using a size-exclusion column as described earlier. Dilute proteoliposomes (20 x) into a chloride-free buffer which has isethionate as a nonpermeable anion to maintain the ionic strength of the extravesicular solution.

3. Add 0.1–0.5 μM lucigenin into the extravesicular buffer.

4. Measure fluorescence by excitation at 455 nm and emission at 505 nm. Continued efflux of Cl$^-$ ions is initiated by adding 1.0 μM valinomycin to shunt charge accumulation due to chloride efflux through the channel (or an anion carrier, cholapod 3 as an example in Figure 6.6b).

5. Once the fluorescence signal is stable, add 5.0 mM detergent (n-decyl-β-D maltopyranoside) to disrupt liposomes and release all chloride ions, reaching a maximal quenching of lucigenin signal.

Figure 6.6b shows an example of using a chloride ionophore in different ratios in vesicles to lead to different degrees of lucigenin quenching. Such an assay is suitable for high-throuput screen of specific modulators of a Cl$^-$ channel.

6.3.6 Activity Assay Using Voltage-Sensitive Probes

Voltage-sensitive dyes can measure the change in transmembrane electrostatic potential. They can be called potential-sensitive probes. These dyes change their

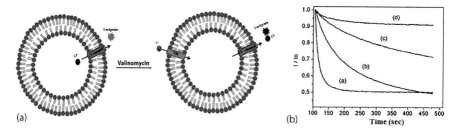

FIGURE 6.6 Schematic diagram showing lucigenin-based Cl⁻ efflux assay. (a) Diagram to show the principle behind the lucigenin-based assay. Purified Cl⁻ channels are reconstituted in lipids vesicles loaded with 400 mM KCl. Vesicles are changed into a Cl⁻ free solution to create a strong gradient for Cl⁻ efflux. Continued chloride efflux is initiated by valinomycin (Vln). Cl⁻ efflux is monitored by quenching of lucigenin. (b) Chloride efflux from vesicles containing cholapod 3 (Cl⁻ ionophore) at a cholapod to lipid ratio of: (a) 1:2,500 (b) 1:25,000 (c) 1:250,000 (d) 1:2,500,000. Na_2SO_4 was outside. (Figure 6.6b reproduced with permission of The Royal Society of Chemistry, London, UK.)

absorption or emission fluorescence as a function of transmembrane potential. This method was started by Lawrence Cohen in the middle of 1970s when he used squid giant axons to screen available dyes,[62–64] and later developed several of them, such as diASP, di-4-ANEPPS,[65–67] di-8-ANEPPS,[68] and di-4-ANEPPDHQ.[69] Due to their chromophores, these dyes are classified as hemocyanin. An array of different potential-sensitive dyes were later developed in order to investigate varying aspects of neuronal functions in real time.

Another group of voltage-sensitive dyes are called ANNINE dyes. ANNINE-6plus is an example. These dyes have better sensitivity and thus an improved signal to noise ratio when compared to the hemocyanin dyes. They can measure even weak subthreshold synaptic potentials, which are generally difficult to do with conventional electrophysiology techniques.[58] By now, a variety of voltage-sensitive dyes are available, differing in signal duration, intensity, signal to noise ratio, and cytotoxicity. A big disadvantage of these dyes is that most of them exhibit a small signal as a fluorescence change (0.1%) over a unit change (say 10 mV) in transmembrane potential, although some may reach 6% with a potential change of 100 mV. Because of such a limitation, low noise level in the detection system is a prerequisite for reliable determination of the relationship between fluorescence change and a change in transmembrane potential.

For better sensitivity, FRET-based voltage-sensitive dyes, such as DiSC3,[5] have been developed. Their response time ranges from 400 μsec to 500 ms and their change in fluorescence intensity can reach 10%–20% per 100 mV.[70] For accurate analysis, double emission ratiometric imaging is usually needed. FLIPR, a commercially available voltage-sensitive dye, has been used to report cell membrane potential. It may reach almost 300% change over background fluorescence when used in neuronal membranes.[71] Roger Tsien's group developed multiple voltage sensitive probes such as VF2.1.Cl, VF2.4.Cl, VF1.4.Cl, etc.[72] One of these VF dyes (VF2.1.Cl) was recently used for measuring potential in purified synaptic vesicles[73,74] (Figure 6.7). Here we will describe the rationale of using FLIPR in vesicles in a stepwise fashion.

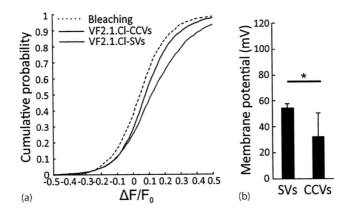

(a) $\Delta F/F_0$ (b)   SVs CCVs

FIGURE 6.7   Sample data showing the use of a voltage-sensitive probe in synaptic vesicles. (a) Cumulative curve generated from $\Delta F/F_0$ in individual VF2.1. Cl-labeled CCVs and SVs in response to 3 mM ATP. The control trace shows the fluorescence change of VF2.1.Cl-labeled CCVs over experimental timescale without ATP addition. (b) Averaged membrane potential in individual VF2.1.Cl-labeled CCVs and SVs upon addition of 3 mM ATP. A smaller membrane potential was recorded across the membrane of SVs in the absence of clathrin coats ($\Delta\psi = 55.02 \pm 2.63$ (SD) in response to 3 mM ATP). (Reproduced from Farsi, Z. et al., *Elife*, 7, e32569, 2018. With permission.)

1. Similar to the use of synaptic vesicles, purified proteins reconstituted in proteoliposomes may be utilized for the same experiments as conducted for synaptic vesicles. Keep salt (K^+ ion) concentration high (~400 mM) during reconstitution.

2. Replace the extravesicular buffer with a K^+-free solution. Dilute the vesicles into the same buffer, which is maintained with the same ionic strength and osmolality as the intra-vesicular solution. Such operation will create a potential difference of approx. −295 mV inside the vesicles, given the average diameter of vesicles is 100 nm.

3. Add 1–5 μL of FLIPR dye, whose amount needs to be optimized by testing with different doses, into the extravesicular buffer. FLIPR is lipophilic.

4. Measure fluorescence by excitation at 530 ± 40 nm and emission at 605 ± 70 nm.[71] Continued flux of K^+ is initiated by adding 1.0–2.0 μM of anion ionophore (chloride ionophore IV to permeate Cl^-; or CCCP to allow H^+ influx), and the vesicular transmembrane potential will decrease to near 0 mv.

5. Once the fluorescence signal is stable, add 5.0 mM detergent (n-decyl-β-D maltopyranoside) to remove potential completely.

The results can be represented as:

$$\frac{\Delta F}{F_0} = \frac{F_t - F_0}{F_0}$$

where F_t = fluorescence measured at given time point (t), F_0 = fluorescence measured at the beginning ($t = 0$).

The measured fluorescence change over time can also be related to the change in membrane potential as

$$\Delta E = \frac{RT}{z'F} \cdot \ln\left(\frac{1}{\Delta F / F_0 + 1}\right)$$

where R and F are the gas and Faraday's constants and T represents the absolute temperature.

In comparison to channel recordings from planar bilayers, the dye-based measurement is easier to implement and suffers no effect from the organic solvent (e.g., decane) used in the planar bilayer system. It is thus possible to apply the assay in high-throughput screen.

6.4 LIMITATIONS OF RECONSTITUTED SYSTEMS IN STUDYING LIPID-PROTEIN INTERACTIONS AND POTENTIAL FUTURE DEVELOPMENTS

6.4.1 STRENGTHS OF RECONSTITUTED SYSTEMS

Reconstituted membrane systems have obvious advantages and shortcomings when compared to native cell membranes. The advantages are reflected in several aspects. (1) Reconstituted membranes are homogeneous in terms of lipid distribution such that usually the lipid composition is similar to or the same as the starting lipids in the protein/lipids/detergents mixture. When a dialysis tubing is used to make proteoliposomes, it is possible to enforce the lipid composition. The BioBeads might have differential interactions with lipids and detergents and incur the deviation of vesicles from the lipid composition of the input mixture, although their pores are more likely to sequester the smaller hydrophobic tails of detergent molecules.[75] (2) With slower removal of detergents, it is possible to achieve higher efficiency of protein reconstitution so that a majority, if not all, of the input membrane proteins are reconstituted into the lipid bilayers. (3) With bSUMs, the anchorage of membrane proteins from one side enables a selective orientation of the proteins in the vesicles, a type of vectorial integration which has not been well controlled in the past. (4) With pure proteins in vesicles, it becomes possible to demonstrate the necessity or sufficiency of the target proteins in fulfilling a specific function when contributions from residual contaminants or even small chemicals can be reliably excluded. It is very difficult to reach such a high level of certainty in a cell-based environment. (5) Reconstituted giant unilamellar vesicles (GUVs) or bSUMs are suitable for patch clamp recordings in single channel levels or for giant patch recordings. With giant patch electrodes sealed on sizeable bSUM vesicles of 20 μm in diameter, it will become feasible to record from ~10^5 channels or transporters in the same time. With a *Shaker*-like Kv channel, a gating current of ~50 pA could be recorded in such conditions. With bSUMs of 50 μm in diameter, a similar level of channel density could be achieved. With vectorial insertion, a bacteriorhodopsin/H⁺-ATPase combination may be used to produce bSUMs that can convert light energy into ATP, a simple artificial photosynthetic system. (6) Lipid vesicles can be prepared with specific, artificial ionic conditions that might not be feasible to achieve with live cells, especially with cytotoxic molecules. Within lipid vesicles, the target proteins are much more stable than those in detergents and can be stored in liquid N_2 for a long time. Proteins in vesicles may be tested against a broad range of parameters that are not feasible for live cells. For example, temperature varying from 0°C to +60°C can be applied to a reconstituted system. Salt concentrations from 0 to 4.0 M can be applied. Faster kinetics may be measured by using lipid vesicles, especially bSUMs, because of better space clamping in a Nanion system and easy operations. (7) Reconstituted lipid vesicles can be used for many other purposes. Membrane fusion assays, stealth vesicles for targeted delivery of anti-cancer drugs or siRNAs,[76] fusion of vesicles into planar lipid bilayers or droplet interface bilayers,[77] tethered bilayers,[76,78] multi-stacked bilayers,[79] etc., all utilize reconstitution to introduce membrane proteins. (8) Reconstituted membranes, especial bSUMs and giant unilamellar vesicles (GUVs), offer a good platform for studying the protein-lipid interaction with precise control of lipid composition and the capability of keeping membranes in a fluidic, homogeneous phase.

Fluorescence-based assays for ion fluxes are suitable for high-throughput screening of specific modulators of transporters or ion channels that allow movement of ions across vesicular membranes. With antimicrobial resistance becoming a daunting problem worldwide, many different ion channels and transporters essential to bacterial survival will be suitable as druggable targets for developing new therapeutic compounds.

6.4.2 SOME LIMITATIONS OF RECONSTITUTED MEMBRANE SYSTEMS

The challenges the reconstituted vesicles face also come from multiple directions. First, proteoliposomes usually do not harbor asymmetry of native cell membranes. Even for cholesterol, recent imaging with cholesterol-specific binding proteins showed that there is a significantly lower level of cholesterol in the inner leaflet than the outer leaflet in the plasma membrane of a live cell.[80] Reconstituted membranes usually lack the lipid transporters that maintain lipid asymmetry. Further, the orientations of different membrane proteins are not easy to control precisely such that the man-made membranes are still not able to reflect the protein-lipid organizations observed in cell membranes. With the strong lipid-dependent gating effects observed on Kv channels, Nav1.8 and Kv4.3 in different cell types,[81,82] it is an inevitable necessity to be cautious when the structural and functional insights into particular membrane proteins in nanodiscs or vesicles are applied to the same proteins in native cell membranes, even though we expect that some of the fundamental biophysical and biochemical properties remain the same despite changes in lipid environments. With the structures and functions of scramblases and other lipid transporters being revealed,[83,84] it will be feasible to reconstitute them into proteoliposomes and generate asymmetrical distribution of lipid molecules.

Second, proteoliposomes may be physically different from cell membranes because the latter have strong support from cytoskeletons and extracellular matrix. Mechanical property and lateral diffusion of proteins and lipids may be altered by these mechanical linkages from both sides. bSUMs have bead-support, but the support is not as dynamic (adaptive) as the cytoskeleton network, and there is still no extracellular matrix present.

Third, proteoliposomes may still be limited in reconstituting multi-component biological vesicles. For example, a synaptic vesicle contains several dozens of different types of membrane proteins that are precisely controlled in copy numbers and in arrangement such that it remains a daunting challenge in order to package these molecules into an artificial synthetic vesicle that would mimic the function of a native synaptic vesicle.[85] It would be a great milestone if bSUMs can be used to reconstitute such a synthetic synaptic vesicle, which would be the first synthetic organelle. With the known components for synaptic vesicles, it will eventually become feasible to do in the future.

Fourth, the proteoliposomes are usually near equilibrium, meaning that they don't have the proton-gradients often reported for intracellular vesicles, such as trafficking vesicles, secretory granules, synaptic vesicles, endosomes, etc. With recombinant vesicular H^+-ATPase and the newly discovered CHGB Cl^- channel or a CLC1/2, it will become feasible to reconstitute vesicles with these proteins such that supplying ATP will drive proton influx and continued acidification of the interior of the proteoliposomes. The energy of the proton gradient will then be utilized to load specific cargo molecules into the vesicles, such as neurotransmitters. Not only the vesicles will be driven out of equilibrium, but also will be properly packed for specific functions.

6.4.3 Perspectives

Even with the latest technologies for quantitative and high-sensitivity analysis of specific lipid molecules,[86,87] it remains impossible to recognize specific lipids in a small patch of membrane. For a patch of cell membrane that is 100 nm in diameter and whose 50% area is occupied by proteins, there are approx. 140,000 lipid molecules. Currently there is no method that is capable of measuring the lipid composition of such a membrane patch or resolving individual lipid molecules around individual membrane proteins on a cell's surface.[88,89] The reconstituted proteoliposomes provide a possible way to make it at least feasible to analyze lipid composition in simpler lipid conditions and study lipid-distribution in the vicinity of a protein molecule. High resolution secondary ion mass spectrometry (SIMS) is currently limited at ~20 nm in resolution. With the focused X-ray beam to become smaller than 1–2 nm and the sensitivity of mass spectrometry (MS) detection can reach the ultimate limit of one ion at a time, it will become feasible to detect individual lipid molecules within 1–2 nm of a protein. If the focused X-ray beam (possibly from a free electron X-ray laser and a proper phase plate) can be made smaller than 1.0 nm, it would become possible to resolve one or two lipid molecules each time, leading to the resolution of individual lipids in the annular layer around a membrane protein.

Alternatively, a focused electron beam may be focused into a diameter as small as 0.5–1.0 nm and a current density of 0.1 A/cm^2, it will deliver ~60 e$_0$/Å^2/s. Such a beam could be used to bombard the surface of a proteoliposome or a cell membrane. With a high-sensitivity detection of vaporized ions, it would be possible to reach high-resolution detection of individual lipid molecules within a membrane patch that is as small as 5 nm in diameter or around a membrane protein.

With highly focused and high-brilliance electron pulse in the subnanosecond range and high-sensitivity detection in electron counting, it will become feasible to collect all diffraction components from a small volume of specimens, say one proteoliposome, and use the signals to deduce the structure of individual lipids in a small patch of membrane. The potential of such a method will be huge if the beam preparation is realized.[90]

6.5   CONCLUSIONS

Reconstituted membrane systems are important tools for studying membrane proteins. As exemplified with ion channels, quantitative analysis of ion channel activity in different reconstituted systems is feasible. Channel activity detected by various assays described in this chapter can be analyzed (semi)-quantitatively and will provide strategies for identifying channel-specific modulators in a high-throughput fashion. Similar assays can be developed for transporters using the same principles described here.

Notes:

1. Weigh 1.0 g of G-50 resin in a 50 mL conical tube and fill the tube with DI water. Soak the resin for 2–3 hours with end-over-end rotation. Add enough hydrated resin into a 5.0 mL empty column (BioRad Inc.) to make a 1.0 mL bed volume. Spin the column at 500 rpm for 1–2 minutes. Wash the column 3–4 times with MilliQ water (1.0 mL) followed by a flux assay buffer twice. Now the column is ready for loading the vesicles.

2. Given that the column is for size-exclusion, it has void volume. Before loading the vesicles, it is recommended to find out the void volume for the column. One can separate blue dextran to do so. Generally, one-third of the bed volume is roughly the void volume for most of the size-exclusion columns that are manually packed.

3. Several other methods can be used to disrupt the multi-lamellar vesicles formed after step 3 of proteoliposomes preparation, with sonication probably the most widely used. Here is a list of possible operations.
   a. French press cell: extrusion
   b. Freeze-thawed liposomes plus extrusion
   c. Lipid film hydration by hand shaking or freeze drying
   d. Microemulsification
   e. Membrane extrusion
   f. Dried reconstituted vesicles

4. By varying the protein: lipid ratio in the proteoliposomes, it is feasible to control the average number of functional channels in each liposome. The following can be used to estimate the average number of liposomes in the solution and hence the approximate number of functional channels per liposome assuming an average size of unilamellar liposomes to be 100 nm in diameter.

$$N_t = \frac{4\Pi r^2 + 4\Pi \left[r - h \right]^2}{a}$$

Here $4\Pi r^2$ represents the outer surface area of the liposome; $r = d/2$, is the radius of the liposome; h the thickness of the bilayer which is approx. 5.0 nm; and a is the average area of the lipid headgroup. The headgroup area of a phosphatidylcholine lipid is ~0.71 nm^2.

The above equation can be simplified for the PC lipids as

$$N_{\text{total}} = 17.69 \times \left[r^2 + \left(r - 5 \right)^2 \right]$$

A 100-nm empty phosphatidylcholine liposome contains approx. 80,000 lipid molecules.

The following equation can be used for estimating liposome concentration at a particular lipid concentration:

$$N_{\text{liposomes}} = \frac{M_l \times N_A}{N_t \times 1000}$$

Here M_l = Molar concentration of lipids; N_A = Avogadro's number = 6.023×10^{23}, and N_t = total number of lipid molecules per liposome.

DECLARATION

The authors claim no conflict of interest.

ACKNOWLEDGMENTS

The research in the Jiang laboratory over the years has been supported by NIH (R21GM131231, R01GM111367, R01GM093271 & R01GM088745), AHA (12IRG9400019), CF Foundation (JIANG15G0), Welch Foundation (I-1684) and CPRIT (RP120474). We thank many colleagues in the ion channel field and in lipid research for their valuable suggestions and advice.

REFERENCES

1. Alenghat, F. J., and Golan, D. E. (2013) Membrane protein dynamics and functional implications in mammalian cells. *Curr Top Membr* **72**, 89–120.
2. Johnson, J. E., and Cornell, R. B. (1999) Amphitropic proteins: Regulation by reversible membrane interactions (review). *Mol Membr Biol* **16**, 217–235.
3. Boyd, D., Schierle, C., and Beckwith, J. (1998) How many membrane proteins are there? *Protein Sci* **7**, 201–205.
4. Krogh, A., Larsson, B., von Heijne, G., and Sonnhammer, E. L. (2001) Predicting transmembrane protein topology with a hidden Markov model: Application to complete genomes. *J Mol Biol* **305**, 567–580.

5. Lander, E. S., Linton, L. M., Birren, B., Nusbaum, C., Zody, M. C., Baldwin, J., Devon, K. et al., and International Human Genome Sequencing, C. (2001) Initial sequencing and analysis of the human genome. *Nature* **409**, 860–921.

6. Wagner, M. L., and Tamm, L. K. (2000) Tethered polymer-supported planar lipid bilayers for reconstitution of integral membrane proteins: Silane-polyethyleneglycol-lipid as a cushion and covalent linker. *Biophys J* **79**, 1400–1414.

7. Drew, D., Froderberg, L., Baars, L., and de Gier, J. W. (2003) Assembly and overexpression of membrane proteins in *Escherichia coli*. *Biochim Biophys Acta* **1610**, 3–10.

8. le Maire, M., Champeil, P., and Moller, J. V. (2000) Interaction of membrane proteins and lipids with solubilizing detergents. *Biochim Biophys Acta* **1508**, 86–111.

9. Garavito, R. M., and Ferguson-Miller, S. (2001) Detergents as tools in membrane biochemistry. *J Biol Chem* **276**, 32403–32406.

10. Kosolapov, A., and Deutsch, C. (2009) Tertiary interactions within the ribosomal exit tunnel. *Nat Struct Mol Biol* **16**, 405–411.

11. Alberts, B., Johnson, A., Lewis, J., Raff, M., Roberts, K., and Walter, P. (2007) *Molecular Biology of the Cell*, 5th ed., Galand Science, New York.

12. Guidotti, G. G., Luneburg, B., and Borghetti, A. F. (1969) Amino acid uptake in isolated chick embryo heart cells. *Biochem J* **114**, 97–105.

13. Palmieri, F., and Klingenberg, M. (1979) Direct methods for measuring metabolite transport and distribution in mitochondria. *Methods Enzymol* **56**, 279–301.

14. Burchell, A. (1996) Endoplasmic reticulum phosphate transport. *Kidney Int* **49**, 953–958.

15. Deriy, L. V., Gomez, E. A., Jacobson, D. A., Wang, X., Hopson, J. A., Liu, X. Y., Zhang, G., Bindokas, V. P., Philipson, L. H., and Nelson, D. J. (2009) The granular chloride channel ClC-3 is permissive for insulin secretion. *Cell Metab* **10**, 316–323.

16. Jentsch, T. J., Maritzen, T., Keating, D. J., Zdebik, A. A., and Thevenod, F. (2010) ClC-3—a granular anion transporter involved in insulin secretion? *Cell Metab* **12**, 307–308.

17. Lee, S., Zheng, H., Shi, L., and Jiang, Q. X. (2013) Reconstitution of a Kv channel into lipid membranes for structural and functional studies. *J Vis Exp* **77**, e50436.

18. Gulik-Krzywicki, T., Seigneuret, M., and Rigaud, J. L. (1987) Monomer-oligomer equilibrium of bacteriorhodopsin in reconstituted proteoliposomes: A freeze-fracture electron microscope study. *J Biol Chem* **262**, 15580–15588.

19. Rigaud, J. L., Bluzat, A., and Buschlen, S. (1983) Incorporation of bacteriorhodopsin into large unilamellar liposomes by reverse phase evaporation. *Biochem Biophys Res Commun* **111**, 373–382.

20. Seigneuret, M., Favre, E., Morrot, G., and Devaux, P. F. (1985) Strong interactions between a spin-labeled cholesterol analog and erythrocyte proteins in the human erythrocyte membrane. *Biochim Biophys Acta* **813**, 174–182.

21. Zheng, H., Lee, S., Llaguno, M. C., and Jiang, Q. X. (2016) bSUM: A bead-supported unilamellar membrane system facilitating unidirectional insertion of membrane proteins into giant vesicles. *J Gen Physiol* **147**, 77–93.

22. Rigaud, J. L., and Levy, D. (2003) Reconstitution of membrane proteins into liposomes. *Methods Enzymol* **372**, 65–86.

23. de Almeida, R. F., Fedorov, A., and Prieto, M. (2003) Sphingomyelin/phosphatidylcholine/cholesterol phase diagram: Boundaries and composition of lipid rafts. *Biophys J* **85**, 2406–2416.

24. Cladera, J., Rigaud, J. L., Villaverde, J., and Dunach, M. (1997) Liposome solubilization and membrane protein reconstitution using Chaps and Chapso. *Eur J Biochem* **243**, 798–804.

25. Yadav, G. P., Zheng, H., Yang, Q., Douma, L. G., Bloom, L. B., and Jiang, Q. X. (2018) Secretory granule protein chromogranin B (CHGB) forms an anion channel in membranes. *Life Sci Alliance* **1**, e201800139.

26. Bonomi, F., and Kurtz, D. M., Jr. (1984) Chromatographic separation of extruded iron-sulfur cores from the apoproteins of *Clostridium pasteurianum* and spinach ferredoxins in aqueous Triton X-100/urea. *Anal Biochem* **142**, 226–231.

27. Lorusso, D. J., and Green, F. A. (1975) Reconstitution of Rh (D) antigen activity from human erythrocyte membranes solubilized by deoxycholate. *Science* **188**, 66–67.

28. Welling, G. W., Nijmeijer, J. R., van der Zee, R., Groen, G., Wilterdink, J. B., and Welling-Wester, S. (1984) Isolation of detergent-extracted Sendai virus proteins by gel-filtration, ion-exchange and reversed-phase high-performance liquid chromatography and the effect on immunological activity. *J Chromatogr* **297**, 101–109.

29. Shechter, Y., Chang, K. J., Jacobs, S., and Cuatrecasas, P. (1979) Modulation of binding and bioactivity of insulin by anti-insulin antibody: Relation to possible role of receptor self-aggregation in hormone action. *Proc Natl Acad Sci U S A* **76**, 2720–2724.

30. Shechter, I., and Bloch, K. (1971) Solubilization and purification of trans-farnesyl pyrophosphate-squalene synthetase. *J Biol Chem* **246**, 7690–7696.

31. Garland, R. C., and Cori, C. F. (1972) Separation of phospholipids from glucose-6-phosphatase by gel chromatography. Specificity of phospholipid reactivation. *Biochemistry* **11**, 4712–4718.

32. Gibson, G. G., and Schenkman, J. B. (1978) Purification and properties of cytochrome P-450 obtained from liver microsomes of untreated rats by lauric acid affinity chromatography. *J Biol Chem* **253**, 5957–5963.

33. Momoi, T. (1979) The presence of lipophilic glycoprotein interacting with insulin. *Biochem Biophys Res Commun* **87**, 541–549.

34. Warner, M. (1982) Catalytic activity of partially purified renal 25-hydroxyvitamin D hydroxylases from vitamin D-deficient and vitamin D-replete rats. *J Biol Chem* **257**, 12995–13000.

35. Rigaud, J. L., Paternostre, M. T., and Bluzat, A. (1988) Mechanisms of membrane protein insertion into liposomes during reconstitution procedures involving the use of detergents. 2. Incorporation of the light-driven proton pump bacteriorhodopsin. *Biochemistry* **27**, 2677–2688.

36. Paternostre, M. T., Roux, M., and Rigaud, J. L. (1988) Mechanisms of membrane protein insertion into liposomes during reconstitution procedures involving the use of detergents. 1. Solubilization of large unilamellar liposomes (prepared by reverse-phase evaporation) by triton X-100, octyl glucoside, and sodium cholate. *Biochemistry* **27**, 2668–2677.

37. Zheng, H., Liu, W., Anderson, L. Y., and Jiang, Q. X. (2011) Lipid-dependent gating of a voltage-gated potassium channel. *Nat Commun* **2**, 250.

38. Iwahashi, Y., and Nakamura, T. (1989) Orientation and reactivity of NADH kinase in proteoliposomes. *J Biochem* **105**, 922–926.

39. Civjan, N. R., Bayburt, T. H., Schuler, M. A., and Sligar, S. G. (2003) Direct solubilization of heterologously expressed membrane proteins by incorporation into nanoscale lipid bilayers. *Biotechniques* **35**, 556–560, 562–553.

40. Denisov, I. G., Grinkova, Y. V., Lazarides, A. A., and Sligar, S. G. (2004) Directed self-assembly of monodisperse phospholipid bilayer nanodiscs with controlled size. *J Am Chem Soc* **126**, 3477–3487.

41. Denisov, I. G., and Sligar, S. G. (2016) Nanodiscs for structural and functional studies of membrane proteins. *Nat Struct Mol Biol* **23**, 481–486.

42. Bayburt, T. H., and Sligar, S. G. (2010) Membrane protein assembly into Nanodiscs. *FEBS Lett* **584**, 1721–1727.

43. Shaw, A. W., McLean, M. A., and Sligar, S. G. (2004) Phospholipid phase transitions in homogeneous nanometer scale bilayer discs. *FEBS Lett* **556**, 260–264.

44. Inagaki, S., Ghirlando, R., and Grisshammer, R. (2013) Biophysical characterization of membrane proteins in nanodiscs. *Methods* **59**, 287–300.

45. Hagn, F., Etzkorn, M., Raschle, T., and Wagner, G. (2013) Optimized phospholipid bilayer nanodiscs facilitate high-resolution structure determination of membrane proteins. *J Am Chem Soc* **135**, 1919–1925.
46. Justesen, B. H., Hansen, R. W., Martens, H. J., Theorin, L., Palmgren, M.G., Martinez, K.L., Pomorski, T.G., and Fuglsang, A.T. (2013) Active plasma membrane P-type H+-ATPase reconstituted into nanodiscs is a monomer. *J Biol Chem* **288**(37), 26419–26429.
47. McConnell, H. M., Watts, T. H., Weis, R. M., and Brian, A. A. (1986) Supported planar membranes in studies of cell-cell recognition in the immune system. *Biochim Biophys Acta* **864**, 95–106.
48. Tamm, L. K. (1988) Lateral diffusion and fluorescence microscope studies on a monoclonal antibody specifically bound to supported phospholipid bilayers. *Biochemistry* **27**, 1450–1457.
49. Urisu, T., Rahman, M. M., Uno, H., Tero, R., and Nonogaki, Y. (2005) Formation of high-resistance supported lipid bilayer on the surface of a silicon substrate with microelectrodes. *Nanomedicine* **1**, 317–322.
50. Dimitrievski, K., Reimhult, E., Kasemo, B., and Zhdanov, V. P. (2004) Simulations of temperature dependence of the formation of a supported lipid bilayer via vesicle adsorption. *Colloids Surf B Biointerfaces* **39**, 77–86.
51. Furukawa, K., Sumitomo, K., Nakashima, H., Kashimura, Y., and Torimitsu, K. (2007) Supported lipid bilayer self-spreading on a nanostructured silicon surface. *Langmuir* **23**, 367–371.
52. Deverall, M. A., Gindl, E., Sinner, E. K., Besir, H., Ruehe, J., Saxton, M. J., and Naumann, C. A. (2005) Membrane lateral mobility obstructed by polymer-tethered lipids studied at the single molecule level. *Biophys J* **88**, 1875–1886.
53. Jin, A. J., Huster, D., Gawrisch, K., and Nossal, R. (1999) Light scattering characterization of extruded lipid vesicles. *Eur Biophys J* **28**, 187–199.
54. Stockbridge, R. B., Robertson, J. L., Kolmakova-Partensky, L., and Miller, C. (2013) A family of fluoride-specific ion channels with dual-topology architecture. *Elife* **2**, e01084.
55. Busto, N., Garcia, B., Leal, J. M., Gaspar, J. F., Martins, C., Boggioni, A., and Secco, F. (2011) ACMA (9-amino-6-chloro-2-methoxy acridine) forms three complexes in the presence of DNA. *Phys Chem Chem Phys* **13**, 19534–19545.
56. Feng, L., Campbell, E. B., and MacKinnon, R. (2012) Molecular mechanism of proton transport in CLC Cl-/H+ exchange transporters. *Proc Natl Acad Sci U S A* **109**, 11699–11704.
57. Pedersen, J. T., Kanashova, T., Dittmar, G., and Palmgren, M. (2018) Isolation of native plasma membrane H(+)-ATPase (Pma1p) in both the active and basal activation states. *FEBS Open Bio* **8**, 774–783.
58. Su, Z., Brown, E. C., Wang, W., and MacKinnon, R. (2016) Novel cell-free high-throughput screening method for pharmacological tools targeting K+ channels. *Proc Natl Acad Sci U S A* **113**, 5748–5753.
59. Krapf, R., Illsley, N. P., Tseng, H. C., and Verkman, A. S. (1988) Structure-activity relationships of chloride-sensitive fluorescent indicators for biological application. *Anal Biochem* **169**, 142–150.
60. Rommens, J. M., Dho, S., Bear, C. E., Kartner, N., Kennedy, D., Riordan, J. R., Tsui, L. C., and Foskett, J. K. (1991) cAMP-inducible chloride conductance in mouse fibroblast lines stably expressing the human cystic fibrosis transmembrane conductance regulator. *Proc Natl Acad Sci U S A* **88**, 7500–7504.
61. Dho, S., and Foskett, J. K. (1993) Optical imaging of Cl-permeabilities in normal and CFTR-expressing mouse L cells. *Biochim Biophys Acta* **1152**, 83–90.
62. Gupta, R. K., Salzberg, B. M., Grinvald, A., Cohen, L. B., Kamino, K., Lesher, S., Boyle, M. B., Waggoner, A. S., and Wang, C. H. (1981) Improvements in optical methods for measuring rapid changes in membrane potential. *J Membr Biol* **58**, 123–137.

63. Ross, W. N., Salzberg, B. M., Cohen, L. B., Grinvald, A., Davila, H. V., Waggoner, A. S., and Wang, C. H. (1977) Changes in absorption, fluorescence, dichroism, and Birefringence in stained giant axons: Optical measurement of membrane potential. *J Membr Biol* **33**, 141–183.

64. Cohen, L. B., Salzberg, B. M., Davila, H. V., Ross, W. N., Landowne, D., Waggoner, A. S., and Wang, C. H. (1974) Changes in axon fluorescence during activity: Molecular probes of membrane potential. *J Membr Biol* **19**, 1–36.

65. Fluhler, E., Burnham, V. G., and Loew, L. M. (1985) Spectra, membrane binding, and potentiometric responses of new charge shift probes. *Biochemistry* **24**, 5749–5755.

66. Loew, L. M., and Lewis, A. (2015) Second harmonic imaging of membrane potential. *Adv Exp Med Biol* **859**, 473–492.

67. Loew, L. M., Cohen, L. B., Dix, J., Fluhler, E. N., Montana, V., Salama, G., and Wu, J. Y. (1992) A naphthyl analog of the aminostyryl pyridinium class of potentiometric membrane dyes shows consistent sensitivity in a variety of tissue, cell, and model membrane preparations. *J Membr Biol* **130**, 1–10.

68. Fisher, J. A., Barchi, J. R., Welle, C. G., Kim, G. H., Kosterin, P., Obaid, A. L., Yodh, A. G., Contreras, D., and Salzberg, B. M. (2008) Two-photon excitation of potentiometric probes enables optical recording of action potentials from mammalian nerve terminals in situ. *J Neurophysiol* **99**, 1545–1553.

69. Obaid, A. L., Loew, L. M., Wuskell, J. P., and Salzberg, B. M. (2004) Novel naphthyl-styryl-pyridium potentiometric dyes offer advantages for neural network analysis. *J Neurosci Methods* **134**, 179–190.

70. Briggman, K. L., Kristan, W. B., Gonzalez, J. E., Kleinfeld, D., and Tsien, R. Y. (2015) Monitoring integrated activity of individual neurons using FRET-based voltage-sensitive dyes. *Adv Exp Med Biol* **859**, 149–169.

71. Fairless, R., Beck, A., Kravchenko, M., Williams, S. K., Wissenbach, U., Diem, R., and Cavalie, A. (2013) Membrane potential measurements of isolated neurons using a voltage-sensitive dye. *PLoS One* **8**, e58260.

72. Miller, E. W., Lin, J. Y., Frady, E. P., Steinbach, P. A., Kristan, W. B., Jr., and Tsien, R. Y. (2012) Optically monitoring voltage in neurons by photo-induced electron transfer through molecular wires. *Proc Natl Acad Sci U S A* **109**, 2114–2119.

73. Milosevic, I. (2018) Revisiting the role of clathrin-mediated endoytosis in synaptic vesicle recycling. *Front Cell Neurosci* **12**, 27.

74. Farsi, Z., Gowrisankaran, S., Krunic, M., Rammner, B., Woehler, A., Lafer, E. M., Mim, C., Jahn, R., and Milosevic, I. (2018) Clathrin coat controls synaptic vesicle acidification by blocking vacuolar ATPase activity. *Elife* **7**, e32569.

75. Rigaud, J. L., Mosser, G., Lacapere, J. J., Olofsson, A., Levy, D., and Ranck, J. L. (1997) Bio-Beads: An efficient strategy for two-dimensional crystallization of membrane proteins. *J Struct Biol* **118**, 226–235.

76. Immordino, M. L., Dosio, F., and Cattel, L. (2006) Stealth liposomes: Review of the basic science, rationale, and clinical applications, existing and potential. *Int J Nanomedicine* **1**, 297–315.

77. Leptihn, S., Castell, O. K., Cronin, B., Lee, E. H., Gross, L. C., Marshall, D. P., Thompson, J. R., Holden, M., and Wallace, M. I. (2013) Constructing droplet interface bilayers from the contact of aqueous droplets in oil. *Nature Protocols* **8**, 1048–1057.

78. Andersson, J., and Koper, I. (2016) Tethered and polymer supported bilayer lipid membranes: Structure and function. *Membranes (Basel)* **6**, 30.

79. Zhu, Y., Negmi, A., and Moran-Mirabal, J. (2015) Multi-stacked supported lipid bilayer micropatterning through polymer stencil lift-off. *Membranes (Basel)* **5**, 385–398.

80. Liu, S. L., Sheng, R., Jung, J. H., Wang, L., Stec, E., O'Connor, M. J., Song, S. et al. (2017) Orthogonal lipid sensors identify transbilayer asymmetry of plasma membrane cholesterol. *Nat Chem Biol* **13**, 268–274.

81. Jiang, Q. X. (2019) Cholesterol-dependent gating effects on ion channels. *Adv Exp Med Biol* **1115**, 167–190.
82. Jiang, Q. X. (2018) Lipid-dependent gating of ion channels. In *Protein-Lipid Interactions: Perspectives, Techniques and Challenges* (Catala, A. ed.), Nova Science Publishers, New York, p. 196.
83. Alvadia, C., Lim, N. K., Clerico Mosina, V., Oostergetel, G. T., Dutzler, R., and Paulino, C. (2019) Cryo-EM structures and functional characterization of the murine lipid scramblase TMEM16F. *eLife* **8**, e44365.
84. Brunner, J. D., Lim, N. K., Schenck, S., Duerst, A., and Dutzler, R. (2014) X-ray structure of a calcium-activated TMEM16 lipid scramblase. *Nature* **516**, 207–212.
85. Gronborg, M., Pavlos, N. J., Brunk, I., Chua, J. J., Munster-Wandowski, A., Riedel, D., Ahnert-Hilger, G., Urlaub, H., and Jahn, R. (2010) Quantitative comparison of glutamatergic and GABAergic synaptic vesicles unveils selectivity for few proteins including MAL2, a novel synaptic vesicle protein. *J Neurosci* **30**, 2–12.
86. Podechard, N., Ducheix, S., Polizzi, A., Lasserre, F., Montagner, A., Legagneux, V., Fouche, E. et al. (2018) Dual extraction of mRNA and lipids from a single biological sample. *Sci Rep* **8**, 7019.
87. Khoury, S., Canlet, C., Lacroix, M. Z., Berdeaux, O., Jouhet, J., and Bertrand-Michel, J. (2018) Quantification of lipids: Model, reality, and compromise. *Biomolecules* **8**, 174.
88. Boxer, S. G., Kraft, M. L., and Weber, P. K. (2009) Advances in imaging secondary ion mass spectrometry for biological samples. *Annu Rev Biophys* **38**, 53–74.
89. Kraft, M. L., Weber, P. K., Longo, M. L., Hutcheon, I. D., and Boxer, S. G. (2006) Phase separation of lipid membranes analyzed with high-resolution secondary ion mass spectrometry. *Science* **313**, 1948–1951.
90. Miao, J., Ercius, P., and Billinge, S. J. (2016) Atomic electron tomography: 3D structures without crystals. *Science* **353**, 6306.

7 Cholesterol Modulation of BK (MaxiK; Slo1) Channels
Mechanistic and Pharmacological Aspects

Alex M. Dopico, Anna N. Bukiya, and Kelsey North

CONTENTS

7.1 INTRODUCTION

In this brief review we present current mechanistic understanding of the interaction(s) between two molecules that are fundamental in mammalian biology: cholesterol (CLR) and Ca^{2+}/voltage-gated K^+ channels of large conductance (BK channels). Nothing will be added here to the plethora of literature documenting the fundamental role of CLR in biology, including its contribution to the structure and function of animal membranes, myelination, and as source of bile acids, vitamin D and steroid hormones. CLR levels in the body require fine adjustment, with a deficit in CLR levels leading to poor evolution from surgery and serving as a negative prognostic factor in critically ill patients (Stachon et al., 2000; Dunham et al., 2003; Guimarães et al., 2008; Vyroubal et al., 2008). In turn, CLR excess (particularly when associated to increased low-density lipoprotein in systemic circulation) constitutes a risk factor for a variety of health disorders, including hypertension, stroke, atherosclerosis and cognitive deficits (Hu et al., 2008; Bui et al., 2009; Granger et al., 2010; Miller et al., 2010).

BK channels are widely expressed in mammalian tissues. In most tissues, BK channels are heteromeric complexes (Figure 7.1) resulting from the association of four BK channel-forming subunits (α), which are encoded by the *KCNMA1* or *Slo1* gene (these homo-tetramers are often refer to as slo1 channels) and small, regulatory subunits (β_1–β_4), which are encoded by *KCNMB1-4* genes. *Slo1* pre-mRNA undergoes abundant alternative splicing, editing and further regulation by miRNA (Shipston and Tian, 2016). These processes, followed by posttranslational modification of both α (Kyle and Braun, 2014; Shipston and Tian, 2016) and β subunits (Lu et al., 2006; Li and Yan, 2016) primarily determine the BK current phenotype. BK channel γ subunits (i.e., leucine-rich repeat containing, or LRRC proteins) have also been identified, and drastically modify BK currents (Lu et al., 2006; Evanson et al., 2014; Li and Yan, 2016) yet the physiological significance of γ subunits in most tissues remains to be fully established.

All BK channels, whether homo- or heterotetramers, have a common phenotype: large conductance (BigK) and exquisite selectivity for K^+ over other monovalents, which are combined with dual gating by voltage and Ca^{2+} (Marty et al., 1984). This basic phenotype is suited by the modular nature of the slo1 protein, which consists of (i) a pore gating domain (PGD), (ii) a voltage-sensor domain (VSD), and (iii) an ion-sensing domain. Both PGD and VSD are of transmembrane (TM) location, including the channel core: S0-S6, in which S1–S6 is highly conserved to the core of TM6 purely voltage-gated K^+ channels. In contrast, the ion-sensing domain, which senses changes in Ca^{2+} and Mg^{2+} within physiological, intracellular levels, is of cytosolic location and usually referred to as cytosolic tail domain (CTD) (Figure 7.1)

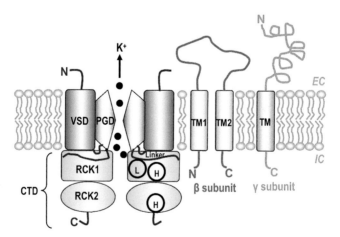

FIGURE 7.1 Schematic structure of BK channel. Functional channel is formed by four α subunits (slo1 protein) depicted in blue. In vertebrates, functional homotetramers are accompanied by small accessory subunits of β and/or γ type, depicted in green and orange, respectively. CRAC4: cholesterol recognition amino acid consensus domain 4; CTD: cytosolic tail domain; EC: extracellular; IC: intracellular; PGD: pore gating domain; RCK: regulator of conductance of potassium (domain); TM: transmembrane; VSD: voltage-sensor domain; Y: central tyrosine within CRAC4 (see main text).

(Wang and Sigworth, 2009; Yuan et al., 2010; Wu et al., 2010; Tao et al., 2017). BK channel's dual gating by depolarization of transmembrane potential and/or increased Ca^{2+} levels near the channel's cytosolic Ca^{2+} sensors (Figure 7.1) allows this channel to link membrane potential to Ca^{2+} signaling and, thus, participate in most major biological processes, including neuronal firing, neurotransmitter and hormonal release, circadian rhythms, myogenic tone, inflammation, immunity, and cancer metastasis (Meredith et al., 2004; Catacuzzeno et al., 2015; Balderas et al., 2015; Griguoli et al., 2016; Latorre et al., 2017; Dopico et al., 2018). Given their importance in normal physiology and pathophysiology, and the fact that CLR and BK channels serve as molecular models for type II lipids (that is, a lipid with cross-sectional area of the headgroup being smaller than the area occupied by the tail) and TM modular proteins respectively, the interaction between CLR and BK channels is a question that extends to the fields of biophysics, biochemistry, cell and membrane biology, physiology and pathology, pharmacology, medicinal chemistry, and therapeutics.

Modulation of BK currents by CLR *in native tissues* may involve a myriad of mechanisms and complex processes, including genetic, epigenetic, membrane-bound and cytosolic factors, modification of cytoarchitecture of proteolipid domains where the BK proteins reside (CLR-enriched lipid rafts in particular), the local ionic medium and, finally, functional interactions between CLR molecules and the proteins that constitute native BK channels (reviewed in Dopico et al., 2012). This mini-review focuses exclusively on the latter, which we will label as direct interactions, in opposition to the more complex mechanisms that require a plethora of signaling molecules, organelles additional to the cell membrane (e.g., endosomes) and/or cell integrity, i.e., indirect interactions. Following a dichotomy first presented to explain the modulation of ion channel function by alcohols and general anesthetics (reviewed by Peoples et al., 1996), CLR-BK channel direct interactions have been interpreted within two major schools of thought: the first (lipid theory), as initially introduced, has favored the idea that modification of BK channel function by CLR is the consequence of CLR-induced changes in the physical properties of the *bulk* lipid bilayer where the BK channel proteins reside, these changes being secondary to CLR molecules' insertion in the bilayer lipid (in particular, within the bilayer hydrophobic core). Later, discovery of membrane lipid domains where the lipid composition and architecture differ from those in the bulk bilayer, in particular CLR-enriched vertical domains termed "rafts" (Brown and London, 1998; Owen and Gaus, 2013; Kim and London, 2015), as well as documentation that membrane spanning proteins can sort membrane lipids in their vicinity (Simons and Sampaio, 2011), leads to further refinement of the lipid theory: CLR alters BK function by modifying the physical properties of the lipids that reside immediately around the TM domains of the channel (i.e., boundary lipids). Of note, the importance of boundary lipids and CLR in ion channel function has been first and significantly developed for the nicotinic acetylcholine receptor (Barrantes, 2004).

In contrast to the lipid theory, the second school of thought on CLR-BK channel direct interactions states that CLR-induced modification of BK channel function results from direct binding between the steroids and the BK channel protein subunits (i.e., protein theory). While these two interpretations have been traditionally presented as mutually exclusive, there is no reason to downplay their cross-talking:

CLR insertion into the membrane boundary lipids could evoke changes in BK channel protein structure leading to a conformation(s) that facilitates CLR binding, the latter process being the actual determinant of change in channel function. Conversely, CLR binding to BK channel protein subunits might change their structure to distinct conformations or stoichiometric arrangements that facilitate CLR insertion into the lipid medium, which leads to modification of lipid domain physical properties and actual change in BK function. Last but not least, lipid- and protein-mediated mechanisms could *independently* contribute to CLR modulation of BK channel function, as found for the oxytocin and cholecystokinin receptors (Gimpl et al., 1997). Next, we briefly describe both lipid and protein theories of CLR-BK channel direct interactions. The minireview finishes with a section on mechanisms (both lipid- and protein-mediated) that lead to CLR-induced modification of BK channel pharmacology.

7.2   LIPID THEORY

The hypothesis that CLR-induced perturbations of bilayer physical properties lead to modification of BK channel function is based on the following premises. First, most studies of CLR modulation of BK currents were conducted at CLR molar fractions found in natural membranes (Wu et al., 2013; Bisen et al., 2016; Simakova et al., 2017; North et al., 2018; earlier studies have been reviewed in Dopico et al., 2012). At these molar fractions (≤ 50 mol% CLR; Gennis, 1989; Sackmann, 1995) CLR presence in the bilayer alters several physical properties of the lipid bilayer; thus, CLR broadens and eventually eliminates the gel-to-liquid crystalline phase transition, decreases the cross sectional area occupied by the phospholipid in the liquid-crystalline state, increases lateral stress and stiffness of the phospholipid monolayer or bilayer in the physiologically relevant fluid phase, increases the modulus of compressibility and mechanical strength, introduces negative monolayer curvature and increases bilayer thickness, and modifies the bilayer electrical dipole in a nonmonotonic fashion (Helfrich, 1973; McIntosh, 1978; Gruner, 1985; Evans and Needham, 1986; Epand and Bottega, 1987; Needham and Nunn, 1990; Nezil and Bloom, 1992; Chen and Rand, 1997; McMullen et al., 1999; McConnell and Radhakrishan, 2003; Starke-Peterkovic et al., 2006; Chong et al., 2009; Shrestha et al., 2017). Second, both model peptides and oligomeric ion channel complexes change function (mainly gating and thus, steady-state activity) when one of the aforementioned bilayer physical properties is modified (Lundbæk and Andersen, 1994; Lundbæk et al., 1996, 2004; Bezrukov, 2000; Helrich et al., 2006; Bruno et al. 2007; Lundbæk, 2008; Lundbæk et al., 2010). Third, CLR modification of BK channel function involves drastic change in channel steady-state activity (reflected as a drop in channel open probability; Po) with minor, if any, modification of unitary conductance (differences on unitary current amplitude data among studies are discussed in Crowley et al., 2003; Dopico et al., 2012). Moreover, inspection of single BK channel recordings following increase in membrane CLR levels, whether in native cell or artificial phospholipid bilayers, reveal that CLR introduces neither flickery intra-burst closures nor subconductance states (Bolotina et al., 1989; Chang et al., 1995; Crowley et al., 2003; Bukiya et al., 2011b; Singh et al., 2012). Thus, it is fair to speculate that the site(s) that accommodates CLR molecules in or around the BK channel is away from the ion conduction pathway. As noted in the introduction, BK channel-forming slo1 subunits are modular proteins, with the PGD and VSD/Ca^{2+}-sensing domains respectively hidden and exposed to membrane lipids (Figure 7.1). Moreover, extensive documentation from pioneering functional work on electrostatic potential and BK gating (Moczydlowski et al., 1985) to recent structural analysis (Jiang, 2019) document the dependence of BK gating on membrane boundary lipid, including CLR. Thus, the fact that CLR acts on BK channel function almost exclusively as a gating modifier is also consistent with the idea that the steroid modifies gating and thus, steady-state activity, by a lipid mechanism.

From the very first reports on CLR modulation of BK channel function in rat aortic myocyte membranes *coincidental* changes in BK steady-activity and bilayer physical properties in response to modification of membrane CLR levels were documented (Bolotina et al., 1989; Bregestovski and Bolotina, 1989). A pioneer work showed that treatment of aortic smooth muscle cells with mevinolin, a CLR-depleting agent, led to a nine-fold increase in the channel open probability (Po) and a nearly two-fold increase in the rotational diffusion coefficient of diphenylhexatriene (DPH) (Bolotina et al., 1989), the latter being an indicator of lipid packing and order in the apolar regions of the membrane (Van Blitterswijk et al., 1981). Conversely, addition of CLR to aortic smooth muscle cell membranes resulted in a two-fold decrease in BK Po and a nearly two-fold decrease in DPH rotational diffusion. These data led authors conclude that modifications in the kinetic properties of BK channels by membrane CLR were "presumably *due* to changes in plasma membrane fluidity" (Bolotina et al., 1989) which, in more recent and accurate terminology, implies an increase in membrane lipid order parameters. Arguments against a major involvement of lipid order/packing in CLR modification of BK function are also indirect: first, while CLR inhibition of BK channels reconstituted into planar phospholipid bilayers shows a monotonic dependence on CLR molar fraction (0–50 mol%; Crowley et al., 2003; Bukiya et al., 2008), CLR distribution and formation of packed complexes with phospholipids into superlattices within the lipid bilayer (Chong et al., 2009) and membrane-associated properties, such as membrane packing, vary with CLR molar fraction in a nonmonotonic manner (Sugár and Chong, 2012). In addition, generalized polarization of Laurdan in large unilamellar vesicles where CLR was probed at small increments within a biologically relevant range (20–50 mol%) shows that fluorescence anisotropy does not change monotonically with increased CLR levels either (Venegas et al., 2007). Second, structure-activity relationship (SAR) data show that CLR analogs (CLR and epiCLR) having similar effect on bulk bilayer order, as measured by fluorescence polarization anisotropy (Gimpl et al., 1997; Xu and London, 2000), drastically differ in their inhibitory efficacy of BK channel activity (Bukiya et al., 2011a). The same study also reveals that coprostanol (COPR), while having an anti-CLR effect on bilayer lipid order (Xu and London, 2000) is an effective inhibitor of BK channel activity (Bukiya et al., 2011a).

Indirect evidence also helped to push the idea that introduction of negative monolayer curvature and increased bilayer lateral stress secondary to cholesterol insertion in between bilayer phospholipids play a role in CLR-induced reduction of BK channel mean open time(s) and eventually steady-state activity. According to the umbrella model, CLR relies on polar phospholipid headgroup's coverage to avoid contact with the aqueous phase (Huang, 2009). Thus, individual CLR molecules are inserted as a wedge between phospholipid molecules (Demel and De Kruyff, 1976; Loura and Prieto, 1997; Hayakawa et al., 1998; Preston Mason et al., 2003). Chang et al. (1995) used Arrhenius plots to calculate the activation energy for BK channel transitions from the open to the closed state(s), which is reduced by CLR. Because the computed lateral stress energy that results from the insertion of CLR as a wedge between the phospholipids is significantly larger than the BK activation energy, authors advanced that a CLR-induced increase in lateral stress favored the BK channel to be deflected back to the closed state, leading to the decrease in mean open time (Chang et al., 1995).

CLR-induced decrease in the open probability (Po) of hslo1 channels reconstituted into 1-pamiltoyl-2-oleoyl-phosphatidylethanolamine (POPE) and 1-pamiltoyl-2-oleoyl-phosphatidylserine (POPS) (3:1 w/w) bilayers is drastically reduced in pure POPE bilayers, a reduction that cannot be explained simply by dilution of negatively charged POPS (Crowley et al., 2003). CLR and POPE are both nonlamellar phase-preferring type II lipids. Upon insertion into a bilayer, these lipids increase lateral pressure in the hydrocarbon chain area of the bilayer and thus, introduce a negative monolayer curvature (Bezrukov, 2000), a deformation known to alter gating of ion channel peptides (Lundbæk et al., 1996). Conceivably, a pure POPE bilayer with a high initial degree of lateral stress could be resistant to further modulation by CLR, which could explain the refractoriness of BK channels to CLR in this bilayer. Alternatively, reduced CLR action on BK channels expressed in pure POPE bilayers might result from the relatively low miscibility of CLR in pure PE (McMullen et al., 1999).

Arguments against increase in bilayer lateral stress as the only or primary mechanism underlying CLR inhibition of BK channel activity also raise from structure-activity relationship (SAR) data (Bukiya et al., 2011a). This study shows: first, the differential efficacies of CLR and related monohydroxysterols to reduce channel activity do not follow molecular area rank; second, the computationally predicted energies for CLR (effective BK inhibitor) and epiCLR (ineffective) to adopt a planar conformation are similar; and third, CLR and COPR both inhibit BK channels, yet these sterols have opposite effects on tight lipid packing and, likely, on lateral stress (Xu and London, 2000). In addition, the entropy changes secondary to BK channel open-closed transitions do not tightly follow predictions should lateral stress be the only contributor to CLR reduction of BK activity (Chang et al., 1995).

Actual measurements with small angle X-ray diffraction and atomic force microscopy in combination with bilayer electrophysiology demonstrate that phospholipid bilayer thickness regulates BK (hslo1) channel mean open time (Yuan et al., 2007), which is reduced by CLR presence in the bilayer (Chang et al., 1995; Crowley et al., 2003). However, several points can be raised against bilayer thickness being the only or primary determinant of BK channel inhibition by CLR. Fist, phosphate-phosphate distances in dimyristoylphosphatidylcholine (DMPC) bilayers that contain epiCLR are intermediate between those of CLR-containing and sterol-free DMPC bilayers (Róg and Pasenkiewicz-Gierula, 2003), yet epiCLR is ineffective in reducing BK Po (Bukiya et al., 2011a). Also, atomic force microscopy of DOPE/sphingomyelin (SPM) membranes in the presence of CLR or ent-CLR, i.e., CLR enantiomer (Westover and Covey, 2004; Belani and Rychnovsky, 2008) show similar images while these enantiomers exert a differential effect on channel activity (Yuan et al., 2011).

Collectively, studies so far have failed to demonstrate that modification of a single physical property of the lipid bilayer is the only or primary mechanism determining CLR-induced modification of BK channel activity. However, the lack of systematic manipulation and actual determination of a bilayer property, and its consequence on CLR modulation of BK activity makes it impossible to rule out a role for lipid bilayer mechanisms in CLR action on these channels. In addition, CLR could modify several physical properties of the lipid bilayer with overlapping (or not) concentration-dependency. Thus, it is likely that the CLR aggregate effect on bilayer bulk behavior is determined by steroid actions on several forces that should be systematically studied in concert. These studies would substantially advance the current state of lipid theories and their involvement in CLR modulation of BK currents.

7.3 PROTEIN THEORY

The first experimental observations of CLR modulation of BK channel function via specific CLR-sensing protein site(s) were obtained in planar lipid bilayers. Incorporation of BK channel-forming α subunits into POPE/POPS (3:1 molar ratio) or DOPE/SPM (3:2 molar ratio) artificial lipid bilayers results in BK currents that are progressively inhibited (i.e., decreased Po) as CLR molar fraction in the lipid mixture is increased from 5–10 to 40–45 mol% (Crowley et al., 2003; Bukiya et al., 2008, 2011a; Yuan et al., 2011). As found with 33 mol% CLR, the same amount of cholestanol, a CLR derivative with *trans* configuration in the steroid A/B ring junction, reduces BK Po in planar binary lipid species bilayers (Bukiya et al., 2011a). This outcome would be consistent with a lipid-mediated mechanism; CLR or cholestanol insertion in bilayers similarly increases phospholipid packing and decreases bilayer permeability (Kodama et al., 2004). However, a decrease in BK Po is also observed upon introduction of 33 mol% COPR into phospholipid mixture (Bukiya et al., 2011a). COPR possesses a *cis* configuration in its steroid A/B ring fusion, and is capable of exerting anti-CLR effects in phospholipid membranes, decreasing phospholipid packing and increasing bilayer permeability (Xu and London, 2000). Moreover, a slight modification of the steroid molecule such as the re-orientation of hydroxyl group into α configuration in (e.g., epiCLR, epicholestenol and epiCOPR) ablates steroid-induced decrease in BK Po (Bukiya et al., 2011a). Finally, ent-CLR at 20 and 33 mol% fails to inhibit BK current following the insertion of BK channel-forming α subunits into POPE/POPS (3:1 molar ratio) bilayers (Bukiya et al., 2011a; Yuan et al., 2011) despite similar ability to modify biophysical properties of phospholipid monolayers (Mannock et al., 2003; Alakoskela et al., 2008; Yuan et al., 2011). Differential modulation of protein function by CLR and ent-CLR generally favors the possibility that CLR-sensing is enabled by a protein binding site that does not bind ent-CLR (Westover and Covey, 2004). Thus, stereospecificity of CLR-driven decrease in BK channel open probability has been attributed to direct recognition of steroid molecule by BK protein sites (Bukiya et al., 2011a; Yuan et al., 2011). An explanation for differential functional consequences of natural versus elegant enantiomeric CLR modulation of ion channel function, however, has been put forth recently with the example of inwardly-rectifying Kir2.2 channels. The latter seem capable of binding both CLR and ent-CLR, yet, binding modes of these steroids differ and therefore, functional response to binding of these steroids is stereospecific (Barbera et al., 2017).

Attempts to unravel the molecular underpinning of CLR interactions with specific regions in BK channel subunits have been largely focused on the mapping and functional probing of CLR recognition amino acid consensus (CRAC) motifs within the BK channel-forming α subunit cloned from rat cerebral artery myocytes (cbv1; gene bank accession number AAP82453; Jaggar et al., 2005). Conventional CRAC motifs are defined by a rather lax amino acid sequence: Leu/Val-(X_{1-5})-Tyr-(X_{1-5})-Lys/Arg, where (X_{1-5}) represents up to five amino acids of any kind. Of note, Tyr is located in central areas of the CRAC domain and thus, termed "central Tyr" (Epand, 2006; Di Scala et al., 2017). Cbv1 isoform amino acid sequence analysis points at ten CRAC motifs, three of which (CRACs 1-3) are located within the protein TM core S0-S6 and seven (CRACs 4-10) reside within the CTD (Singh et al., 2012) (Figure 7.1). Notably, removal of CTD via cbv1-coding nucleotide sequence truncation renders BK currents that are insensitive to CLR presence in POPE:POPS (3:1 molar ratio) planar bilayers (Singh et al., 2012). While these data underscore the importance of the CTD for BK channel-forming proteins to sense CLR, it remains unclear whether CTD CRACs, or any other sequence in the CTD for that matter, actually provide binding areas for CLR molecules and/or participate in overall changes in conformation of the BK protein that confer CLR sensitivity and are secondary to CLR binding to TM regions. However, investigations on the possible contribution of CRAC motifs in the cbv1 CTD to the overall CLR sensitivity of BK currents following channel reconstitution in planar lipid bilayers did yield several interesting observations. First, substitution of central Tyr450 with Phe within membrane-adjacent CRAC4 results in partial, yet statistically significant reduction in CLR sensitivity of BK currents. Second, Tyr450Phe substitution in a cbv1 protein that is truncated immediately after CRAC4 leads to total loss of CLR sensitivity. Lastly, cumulative substitution of central Tyr within CRACs 5-10 leads to a progressive loss of CLR sensitivity of BK currents in full-length cbv1 protein (Singh et al., 2012). These data allow us to conclude that while the membrane-adjacent CRAC4 confers CLR sensitivity to BK currents in absence of other CRACs, the presence of multiple CRAC motifs within the CTD sequence leads to a partial recovery of CLR sensitivity in absence of CRAC4. Therefore, it appears that the CTD CRAC motifs can substitute each other in contributing to the CLR sensitivity of BK channels.

The question on the ability of CRAC4 motif to provide actual binding areas to CLR molecules has not been addressed with either structural methods, such as cryogenic electron microscopy, nuclear magnetic resonance, crystallography or biochemical binding. However, computational molecular dynamics simulations demonstrate that CLR molecule can be successfully retained by the cbv1 CRAC4 motif (Singh et al., 2012). While the critical role of cbv1 CTD CRAC motifs in CLR modulation of BK currents is well documented (Singh et al., 2012), CRAC4 and other CTD CRACs may not be the sole providers of CLR-sensitivity to BK channels. Indeed, a recent analysis of crystallographic structures of water-soluble proteins in complex with CLR fails to demonstrate any correlation between the

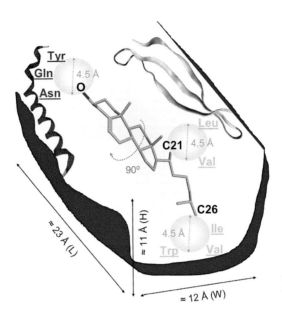

FIGURE 7.2 Common design of cholesterol-binding sites in soluble proteins. Cholesterol (CLR) molecule (teal) is often found between alpha helixes (red) and beta-sheets (yellow). Polar amino acids that are commonly found within CLR- and CLR-sulfate-biding sites are highlighted in red, while aliphatic amino acids are in yellow. Spheres in gray depict the cut-off distance at which steroid-binding amino acids where determined. This research was originally published in the *Journal of Lipid Research*. (From Bukiya, A.N. and Dopico, A.M., *J. Lipid Res.*, 58, 1044–1054, 2017. © The American Society for Biochemistry and Molecular Biology.)

presence of CRAC motifs in the protein and its ability to bind CLR (Bukiya and Dopico, 2017). Based on the analysis of CLR binding sites in these soluble proteins, however, several criteria for CLR binding areas could be developed (Figure 7.2). First, binding areas of hydrophobic CLR in water-soluble proteins must be shielded from aqueous solution, this shielding is provided by a well-structured protein wrap. Second, CLR molecule is likely to reside between α helical and β-sheet secondary structural elements. Third, CLR interaction with individual amino acids at its binding sites points at Asn, Gln and Tyr to most likely participate in hydrogen bonding with CLR hydroxyl group. In turn, Leu, Ile, Val, and Phe at the binding site interact with CLR molecules via hydrophobic interactions (Bukiya and Dopico, 2017). With these three major criteria in mind, we investigated CLR docking on a cbv1 CTD homology model built on a previously published hslo1 CTD crystallographic template (Protein Data Bank accession code 3MT5; Yuan et al., 2010). Site finder suite in Molecular Operating Environment software (MOE by Chemical Computing Group, Canada) reveals 44 potential binding sites for small ligands in the cbv1 CTD. Manual screen of these sites for satisfaction of CLR binding site criteria outlined above, however, reveals one potential lead, which corresponds to a membrane-adjacent protein area (Figure 7.3a). CLR docking onto this site places CLR hydroxyl

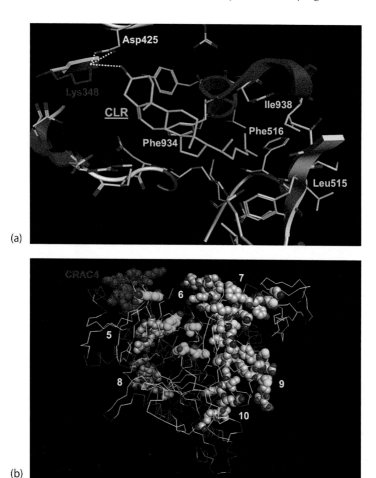

FIGURE 7.3 Cholesterol-sensing sites in BK channel cytosolic tail domain (CTD). (a) Computational analysis of cbv1 CTD structure (see main text) reveals putative CLR-binding site that satisfies the criteria that were identified during the analysis of crystallographic data on CLR molecules bound to water-soluble proteins (Figure 7.2; Bukiya and Dopico, 2017). CLR molecule is depicted in teal. Hydrogen bonds between CLR hydroxyl group and carbon atom 3 and cbv1 amino acids Lys348 and Asp425 are highlighted by white dotted lines. Amino acids that are expected to provide hydrophobic interactions with CLR molecule are in yellow. (b) Overall structure of cbv1 CTD built as homology model on hslo1 CTD template (Protein Data Bank accession code 3MT5; Yuan, P. et al., *Science*, 329, 182–186, 2010). Atoms in amino acids that form cholesterol recognition amino acid consensus (CRAC) motifs are mapped as presented in a ball-mode. CRAC4 is highlighted in pink. Remaining CRACs (CRACs5-10) are sequentially numbered by their first appearance within cbv1 amino acid sequence (Singh, A.K. et al., *J. Biol. Chem.*, 287, 20509–20521, 2012). Membrane bilayer is expected to be located immediately above CRAC4 and CRAC7. Putative CLR-binding site in cbv1 CTD (Figure 7.3a) is adjacent to CRAC4 and is formed by amino acids presented in teal.

group in close proximity to Asp425 and Lys348, these amino acids could serve as hydrogen bonding partners. In turn, CLR steroid core and lateral hydrocarbon chain are abundantly surrounded by Phe, Leu, and Ile amino acids (Figure 7.3a). The type of protein-protein interaction(s) between this putative CLR binding site and the cbv1 CRAC motifs remains speculation. Noteworthy, the BK channel CTD is of cytosolic location. CLR binding sites in water-soluble proteins are often provided by protein tunnels that ensure isolation of hydrophobic CLR molecules from the aqueous media (Bukiya and Dopico, 2017). Maintenance of large hollow tunnels within a globular protein may be energetically unfavorable, thus CLR binding sites in water-soluble proteins are malleable and unravel upon CLR arrival (reviewed in Bukiya and Dopico, 2017). Thus, it is tempting to speculate that CLR sequestering from the membrane may be initially performed by membrane-adjacent CRAC motif(s), such as CRAC4. Such sequestering, in turn, could initiate conformational perturbations that would allow CLR repositioning from CRAC4 to the postulated binding site (Figure 7.3b). Of note, recent computational study mapped a CLR-docking site within the human BK channel voltage-sensing TM area (Wheeler et al., 2019). In this model. CLR forms hydrogen bonds with Lys211 and Asn265. Steroid nucleus is stabilized by interaction with Tyr263, while Ley226 provides hydrophobic support for CLR tail (Wheeler et al., 2019). This newly proposed TM CLR-binding area is upstream of intracellular region that satisfies the criteria for CLR-binding site in water-soluble proteins. Thus, it is possible that TM, CRAC4, and CLR-binding site in CTD work as a bundle to recruit CLR from membrane, sequester it in the membrane-adjacent area and then shuttle it further down to the CTD, respectively. Experimental validation of this model remains to be conducted.

7.4 CHOLESTEROL MODULATION OF BK CHANNEL PHARMACOLOGY: ALCOHOL SENSITIVITY

One of the important effects of CLR on receptor and ion channel function is modification of protein's pharmacological profile. Regarding BK channels, CLR has been reported to modify the channel-forming α, slo1 subunit responses to alcohol (ethyl alcohol; ethanol) (Crowley et al., 2003; Bukiya et al., 2014), as described below.

7.4.1 IN VIVO OBSERVATIONS

Data on CLR modulation of BK channel pharmacology at the most integrative, *in vivo* level, were obtained from studies on CLR modification of BK channel responses to ethanol that impact on cerebral artery diameter. These studies were fueled by the fact that elevated alcohol levels, as well as elevated CLR levels, are known to be independent risk factors for cerebrovascular disease. Considering that more than 86% of adults in the United States consume alcohol (https://www.niaaa.nih.gov/alcohol-health/overview-alcohol-consumption/alcohol-facts-and-statistics), arguably one of the most important drug interactions to consider is ethanol-CLR on BK channels.

Moderate-to-heavy episodic alcohol consumption, such as in binge drinking, substantially alters cerebral artery diameter (Altura and Altura, 1984; Liu et al., 2004; Bukiya et al., 2011b, 2014). The majority of binge drinking in the United States has been reported in the male population (Wilsnack et al., 2009; Fuentes et al., 2017). Of note, an *in vivo* male Sprague Dawley rat model of high (2%) CLR dietary consumption shows that 18–23 weeks of such a diet result in increased blood CLR level and diminished ethanol-induced constriction of pial arterioles in a closed cranial window setting on anesthetized animals (Bukiya et al., 2014). Notably, blood alcohol levels (BAL) following a 50 mM ethanol bolus injection into rat carotid artery remain similar in rats fed high CLR diet and control (iso-caloric) chow (Bukiya et al., 2014). Thus, the difference in ethanol-induced constriction of cerebral arterioles does not stem from differences in ethanol clearance rate between control and experimental groups.

7.4.2 IN VITRO OBSERVATIONS

The results on cerebral arteriole diameter from *in vivo* observations are paralleled by data from *in vitro* studies that use isolated, pressurized (60 mmHg) middle cerebral artery (MCA) segments from Sprague Dawley rats. Indeed, pressurized artery probing with 50 mM ethanol results in decreased ethanol-induced constriction of arteries from rats that are fed high CLR diet when compared to control group (Bukiya et al., 2014). The ethanol concentrations used to probe CLR-ethanol interactions is detected in blood of humans following moderate-to-heavy drinking (Diamond, 1992).

Remarkably, ethanol-induced constriction is restored when rats are receiving high CLR diet supplemented by a daily dose of 10 mg/kg atorvastatin (Simakova et al., 2017), with the statin acting to lower the CLR levels in the high-fat diet-fed rats back to normal. Dietary administration of high CLR or high CLR supplemented with atorvastatin is expected to modify tissue CLR levels. Indeed, biochemical quantification of CLR content and confocal imaging of vascular smooth muscle layer following labeling with sterol-sensitive dye filipin point at increased and normal CLR levels within cerebral artery extra-endothelial compartments following administration of high CLR diet and high CLR diet with atorvastatin, respectively (Bukiya et al., 2014; Simakova et al., 2017). Second, *in vitro* manipulations of CLR levels in de-endothelialized cerebral arteries using CLR carrier methyl-β-cyclodextrin (MβCD; Zidovetzki and Levitan, 2007) are able to modulate ethanol-induced constriction similarly to manipulation of CLR levels using *in vivo* tools (high CLR diet and atorvastatin therapy). In particular, CLR *in vitro* depletion using MβCD restores ethanol-induced constriction of cerebral arteries harvested from rats that are subjected to high CLR diet (Bukiya et al., 2014). Conversely, CLR *in vitro* enrichment using MβCD complex with CLR ablates ethanol-induced constriction observed in cerebral arteries harvested from rats that are subjected to high CLR diet with daily supplementation of atorvastatin (Simakova et al., 2017). Moreover, CLR *in vitro* enrichment in cerebral arteries harvested from rat and mouse species raised on conventional rodent chow ablates ethanol-induced constriction (Bisen et al., 2016).

A similar effect is achieved by CLR *in vitro* enrichment in rat cerebral arteries using low-density lipoproteins (Bisen et al., 2016). Thus, the protective effect of elevated CLR against ethanol-induced constriction is not related to the choice of CLR carrier.

Further *in vitro* studies on rat MCA's following mechanical removal of endothelium reveal that ethanol-induced constriction in a group receiving high CLR diet is ablated when compared to controls (Bukiya et al., 2014). Also, this ethanol-induced constriction is similar in controls and rat donors that received high CLR diet with daily supplementation of atorvastatin. These findings lead to the conclusion that the modulatory effect of high CLR diet and the opposing effect of atorvastatin therapy on ethanol-induced constriction of cerebral arteries do not require a functional endothelium (Simakova et al., 2017). Considering that the majority of the cellular content in cerebral arteries following endothelium removal is provided by smooth muscle cells (Lee, 1995), efforts in identifying the molecular mechanisms that underlie CLR-ethanol interactions were focused on arterial myocytes.

Noteworthy, modification of artery diameter sensitivity to ethanol by elevated CLR levels is not monotonic; CLR *depletion* from rat and mouse cerebral arteries using MβCD also ablates ethanol-induced constriction, albeit at a larger extend than the CLR *in vitro* enrichment (Bukiya et al., 2011b). Thus, both CLR levels above and below those naturally found in de-endothelialized cerebral arteries drastically reduce ethanol-induced cerebrovascular constriction. This bell-shaped dependence of ethanol effect on CLR levels is suggestive of multiple points of regulation that enable CLR-ethanol interactions in controlling cerebral artery diameter. In addition, moderate changes in the CLR levels of cerebral arteries appear sufficient to alter the vessel's ethanol sensitivity. Indeed, in several instances, reported degree of CLR *in vitro* enrichment is modest, not exceeding 10%–20% increase in overall tissue CLR content while ethanol-induced constriction is still very sensitive to this modification in CLR level (Bisen et al., 2016; Simakova et al., 2017). Collectively, data summarized in this paragraph seem to indicate that there is a third-party molecular player that amplifies modest changes in CLR content and/or sensitizes ethanol-induced constriction to CLR modifications (Simakova et al., 2017).

Comparisons of ethanol-induced constriction of cerebral arteries at various ethanol concentrations reveal the competitive nature of CLR-ethanol interaction on cerebral artery diameter (North et al., 2018). Indeed, ethanol-induced constriction is similar in control versus CLR-enriched cerebral arteries when ethanol is probed at the concentration that evokes maximal constriction (100 mM; North et al., 2018). Data obtained at this concentration, which is usually lethal to humans if reached in systemic blood (Tormey and Moore, 2013), are useful for interpreting data in mechanistic studies of CLR-ethanol interactions, as described below.

7.4.3 MECHANISTIC STUDIES; USE OF **CLR** DERIVATIVES AND ROLE OF **BK** CHANNEL SUBUNITS

BK channel subunits are major *functional* targets of ethanol in cerebral arteries; ethanol-induced inhibition of BK channel activity in vascular smooth muscle drives artery constriction in presence of clinically relevant ethanol concentrations

(18–100 mM) (Liu et al., 2004). Remarkably BK currents are significantly inhibited by 50 mM ethanol in membrane patches excised from freshly isolated cerebral artery myocytes from rats on high CLR diet supplemented with daily atorvastatin, whereas such inhibition is lost when BK current is recorded in cell-free patches from rat donors subjected to high CLR diet supplemented by placebo (Simakova et al., 2017). When the former myocytes are subjected to CLR *in vitro* enrichment using MβCD complexed with CLR previously to ethanol *in vitro* administration, ethanol action on BK channels is lost (Simakova et al., 2017). Considering that the use of excised membrane patches makes it highly unlikely that intracellular messengers' action contributes to BK current, the patch-clamp results lead to the conclusion that atorvastatin administration *in vivo* and CLR enrichment of myocytes *in vitro* have the ability to change BK channel pharmacological properties (such as ethanol sensitivity) at the level of the channel lipid microenvironment or the BK channel protein subunits themselves.

In native cellular membranes, including vascular myocytes, BK channels tend to reside within CLR-rich lipid rafts (Dopico and Tigyi, 2007; Sones et al., 2010; Dopico et al., 2012). Modification of CLR levels is expected to alter membrane microdomains, including CLR-rich lipid domains, and sphingolipid distribution (Dopico and Tigyi, 2007; Lingwood, 2011). Thus, one of the possible explanations for CLR-ethanol interaction in controlling BK channel activity arises from CLR modification of lipid rafts. However, CLR modification of BK channel ethanol sensitivity is observed in artificial bilayer systems (see below) of lipid composition that do not enable raft formation (Crowley et al., 2003; Bukiya et al., 2011b). These findings suggest that raft disruption is not a main contributor to CLR modification of the BK channel's ethanol sensitivity.

Changes in membrane CLR levels result in modification of membrane physical properties (see first section), alterations in BK channel's basal Po and its response to ethanol (Crowley et al., 2003; Treistman and Martin, 2009). Indeed, alterations in the acyl chain length, and composition of the membrane lipids modify the channels sensitivity to ethanol in artificial lipid bilayers (Yuan et al., 2008; Treistman and Martin, 2009). To distinguish between membrane- and protein-mediated mechanism of CLR-ethanol interaction in vascular myocytes, recent report utilized CLR derivate COPR. Unlike CLR, COPR possesses saturated A/B ring fusion that results in *cis* configuration of the steroid nucleus (Pine, 1987). As a result, there are striking differences in the abilities of COPR and CLR to modify several major bilayer physical properties (Xu and London, 2000; Le Goff et al., 2007). If the BK channel's ethanol sensitivity is solely determined by the composition of the membrane lipid environment, then it is expected that COPR would increase the channel's sensitivity to ethanol. However, COPR has a protective effect against ethanol-induced constriction of cerebral arteries that is similar to that of CLR (North et al., 2018). The similar effects of COPR and CLR despite their chemical structural differences do not support the idea that alterations of the physical properties of lipid membrane serve as a major contributor to the CLR-ethanol interaction, although such contribution cannot be ruled out. Most likely scenario, however, favors the involvement of a non-selective steroid-sensing protein site within BK channel-forming or accompanying protein(s) that may be present in excised patches. The lack of steroid preference in this protein site is further supported by the data showing that ent-CLR protects against ethanol-induced BK channel inhibition in cerebral artery myocytes and cerebral artery constriction with efficiencies that are similar to that of CLR (Bisen et al., 2016).

Recent findings, however, show that ent-CLR fails to modulate the BK channel's ethanol sensitivity in artificial lipid bilayers formed by DOPE/SPM lipids (Yuan et al., 2011). In this bilayer type, low CLR levels (20% molar fraction) support BK channel activation by ethanol while high levels of CLR (>40% molar ratio) lead to BK channel inhibition by 50 mM ethanol (Yuan et al., 2011). The differential outcomes regarding ent-CLR ability to modify BK channel ethanol sensitivity may arise from differences in (a) lipid composition between vascular myocyte and artificial lipid bilayer, (b) amino acid sequence between pore-forming hslo1 (Yuan et al., 2011) and rat cerebral vessel slo1 (cbv1; Jaggar et al., 2005), and (c) protein environment around the BK channel-forming slo1 proteins. Indeed, while the majority of studies on artificial lipid bilayers engage homomeric BK hslo1, BK channels in vascular myocytes are accompanied by accessory subunits (Brenner et al., 2000; Evanson et al., 2014; Dopico et al., 2018).

To investigate the possible contribution of BK channel subunit composition to BK currents' ethanol sensitivity, our group used the $KCNMB1^{-/-}$ mouse line, that is, a homozygous knock-out for the gene that encodes BK β_1 subunits (Brenner et al., 2000). While ethanol-induced constriction of cerebral arteries and ethanol-induced BK channel inhibition is largely absent in β_1 subunit-free channels (Bukiya et al., 2009), CLR *in vitro* enrichment still counteracts the remnant ethanol sensitivity of BK channels and cerebral arteries from $KCNMB1^{-/-}$ mice by promoting BK channel activation and artery dilation, respectively (Bisen et al., 2016; North et al., 2018). Thus, CLR-ethanol interaction does not require presence of BK β_1 subunits.

Further advances in our understanding of the molecular determinants in CLR-ethanol interactions that govern the BK channel's ethanol sensitivity have emerged recently from mutagenesis and computational modeling studies of the BK channel CRAC4 motif (Figure 7.4) (North et al., 2018). In this work, loading of cerebral arteries from $KCNMA1^{-/-}$ mouse (Meredith et al., 2004) with Slo1 Tyr450Phe-coding plasmid yields BK channels that fail to confer CLR enrichment-dependent protection against ethanol-induced constriction despite the fact that similar procedure with plasmid coding for wild type cbv1 fully mimics CLR-ethanol interaction phenotype observed in rat and wild type mouse arteries (North et al., 2018). Surprisingly, COPR *in vitro* enrichment still ablates ethanol-induced constriction of cerebral arteries loaded with cbv1 Tyr450Phe (North et al., 2018). A plausible explanation for the difference in CLR and COPR experimental outcomes is the inability of COPR to adopt a docking mode that is similar to CLR docking on Tyr450-containing CRAC4 motif. Thus, while protection against ethanol-induced vasoconstriction by elevated CLR requires Tyr450, the COPR data strongly suggest that there are additional mechanisms of steroid-ethanol interaction in control of cerebral artery diameter (North et al., 2018).

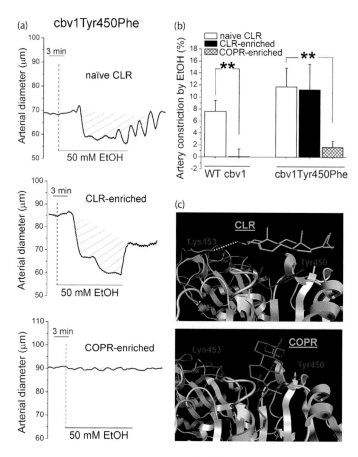

FIGURE 7.4 Mutation of CLR-sensing CRAC4 site at Tyr450 within BK pore-forming (cbv1) subunit fully ablates CLR protection against alcohol-induced vasoconstriction in mouse cerebral artery. (a) Original traces showing alcohol-induced constriction of de-endothelialized MCA from *KCNMA1* KO mouse on C57BL/6 background following permeabilization and introduction of cbv1 with Tyr450Phe mutation. Alcohol was tested on arteries with naïve CLR level (top trace) and arteries that were subjected to *in vitro* enrichment with either CLR (middle) or COPR (bottom). (b) Constriction by 50 mM alcohol with naïve and modified membrane steroid content in *KCNMA1* K/O mouse MCAs following permeabilization with either wild-type cbv1 (WT, first set of bars) or cbv1Tyr450Phe (second set of bars). For WT cbv1, naïve and CLR-enriched groups contained seven arteries each. **Different from naïve CLR (P = 0.0047 by unpaired t-test). For cbv1Tyr450Phe, naïve CLR group contained seven arteries, while CLR- and COPR-enriched groups contained five and eight arteries, respectively. **Different from naïve CLR (P = 0.004 by Mann-Whitney test). (c) Computational docking onto CRACR4-forming amino acids reveals differential docking modes for CLR (top, teal) versus COPR (bottom, orange) molecules. Light dotted line depicts presumed hydrogen bonding between CLR hydroxyl group and Lys453. (With modifications from North, K. et al., *J. Pharmacol. Exp. Ther.*, 367, 234–244, 2018.)

ACKNOWLEDGMENTS

Supported by NIH grants R37 AA11560, R01 HL147315 (AMD) and R01 AA023764 (ANB).

REFERENCES

Alakoskela JM, Sabatini K, Jiang X, Laitala V, Covey DF, Kinnunen PK. 2008. Enantiospecific interactions between cholesterol and phospholipids. *Langmuir* 24: 830–6.

Altura BM, Altura BT. 1984. Alcohol, the cerebral circulation and strokes. *Alcohol* 1: 325–31.

Balderas E, Zhang J, Stefani E, Toro L. 2015. Mitochondrial BK_{Ca} channel. *Front Physiol* 6: 104.

Barbera N, Ayee MA, Akpa BS, Levitan I. 2017. Differential effects of sterols on ion channels: Stereospecific binding vs stereospecific response. *Curr Top Membr* 80: 25–50.

Barrantes FJ. 2004. Structural basis for lipid modulation of nicotinic acetylcholine receptor function. *Brain Res Brain Res Rev* 47: 71–95.

Belani JD, Rychnovsky SD. 2008. A concise synthesis of ent-Cholesterol. *J Org Chem* 73: 2768–73.

Bezrukov S. 2000. Functional consequences of lipid packing stress. *Curr Opin Colloid Interface Sci* 5: 237–43.

Bisen S, Seleverstov O, Belani J, Rychnovsky S, Dopico AM, Bukiya AN. 2016. Distinct mechanisms underlying cholesterol protection against alcohol-induced BK channel inhibition and resulting vasoconstriction. *Biochim Biophys Acta* 1861: 1756–66.

Bolotina V, Omelyanenko V, Heyes B, Ryan U, Bregestovski P. 1989. Variations of membrane cholesterol alter the kinetics of Ca^{2+}-dependent K^+ channels and membrane fluidity in vascular smooth muscle cells. *Pflugers Arch* 415: 262–8.

Bregestovski PD, Bolotina VN. 1989. Membrane fluidity and kinetics of Ca^{2+}-dependent potassium channels. *Biomed Biochim Acta*. 48: S382–7.

Brenner R, Peréz GJ, Bonev AD, Eckman DM, Kosek JC, Wiler SW, Patterson AJ, Nelson MT, Aldrich RW. 2000. Vasoregulation by the beta1 subunit of the calcium-activated potassium channel. *Nature* 407: 870–6.

Brown DA, London E. 1998. Functions of lipid rafts in biological membranes. *Annu Rev Cell Dev Biol* 14: 111–36.

Bruno MJ, Koeppe RE, Andersen OS. 2007. Docosahexaenoic acid alters bilayer elastic properties. *Proc Natl Acad Sci USA* 104: 9638–43.

Bui QT, Prempeh M, Wilensky RL. 2009. Atherosclerotic plaque development. *Int J Biochem Cell Biol* 41: 2109–13.

Bukiya A, Dopico AM, Leffler CW, Fedinec A. 2014. Dietary cholesterol protects against alcohol-induced cerebral artery constriction. *Alcohol Clin Exp Res* 38: 1216–26.

Bukiya AN, Belani JD, Rychnovsky S, Dopico AM. 2011a. Specificity of cholesterol and analogs to modulate BK channels points to direct sterol-channel protein interactions. *J Gen Physiol* 137: 93–110.

Bukiya AN, Dopico AM. 2017. Common structural features of cholesterol binding sites in crystallized soluble proteins. *J Lipid Res* 58: 1044–54.

Bukiya AN, Liu J, Dopico AM. 2009. The BK channel accessory beta1 subunit determines alcohol-induced cerebrovascular constriction. *FEBS Lett* 583: 2779–84.

Bukiya AN, Vaithianathan T, Kuntamallappanavar G, Asuncion-Chin M, Dopico AM. 2011b. Smooth muscle cholesterol enables BK β1 subunit-mediated channel inhibition and subsequent vasoconstriction evoked by alcohol. *Arterioscler Thromb Vasc Biol* 31: 2410–23.

Bukiya AN, Vaithianathan T, Toro L, Dopico AM. 2008. The second transmembrane domain of the large conductance, voltage- and calcium-gated potassium channel beta(1) subunit is a lithocholate sensor. *FEBS Lett* 582: 673–8.

Catacuzzeno L, Caramia M, Sforna L, Belia S, Guglielmi L, D'Adamo MC, Franciolini F. 2015. Reconciling the discrepancies on the involvement of large-conductance Ca$^{(2+)}$-activated K channels in glioblastoma cell migration. *Front Cell Neurosci* 9: 152.

Chang HM, Reitstetter R, Mason RP, Gruener R. 1995. Attenuation of channel kinetics and conductance by cholesterol: An interpretation using structural stress as a unifying concept. *J Membr Biol* 143: 51–63.

Chen Z, Rand RP. 1997. The influence of cholesterol on phospholipid membrane curvature and bending elasticity. *Biophys J* 73: 267–76.

Chong PL, Zhu W, Venegas B. 2009. On the lateral structure of model membranes containing cholesterol. *Biochim Biophys Acta* 1788: 2–11.

Crowley JJ, Treistman SN, Dopico AM. 2003. Cholesterol antagonizes ethanol potentiation of human brain BK$_{Ca}$ channels reconstituted into phospholipid bilayers. *Mol Pharmacol* 64: 365–72.

Demel RA, De Kruyff B. 1976. The function of sterols in membranes. *Biochim Biophys Acta* 457: 109–32.

Di Scala C, Baier CJ, Evans LS, Williamson PTF, Fantini J, Barrantes FJ. 2017. Relevance of CARC and CRAC cholesterol-recognition motifs in the nicotinic acetylcholine receptor and other membrane-bound receptors. *Curr Top Membr* 80: 3–23.

Diamond I. 1992. *Cecil Textbook of Medicine*. Philadelphia, PA: Saunders, pp. 44–47.

Dopico AM, Bukiya AN, Jaggar JH. 2018. Calcium- and voltage-gated BK channels in vascular smooth muscle. *Pflugers Arch* 470: 1271–89.

Dopico AM, Bukiya AN, Singh AK. 2012. Large conductance, calcium- and voltage-gated potassium (BK) channels: Regulation by cholesterol. *Pharmacol Ther* 135: 133–50.

Dopico AM, Tigyi GJ. 2007. A glance at the structural and functional diversity of membrane lipids. *Methods Mol Biol* 400: 1–13.

Dunham M, Fealk M, Sever W. 2003. Following severe injury, hypocholesterolemia improves with convalescence but persists with organ failure or onset of infection. *Critical Care* 7: R145–53.

Epand RM, Bottega R. 1987. Modulation of the phase transition behavior of phosphatidylethanolamine by cholesterol and oxysterols. *Biochemistry* 26: 1820–5.

Epand RM. 2006. Cholesterol and the interaction of proteins with membrane domains. *Prog Lipid Res* 45: 279–94.

Evans E, Needham D. 1986. Giant vesicle bilayers composed of mixtures of lipids, cholesterol and polypeptides. Thermomechanical and (mutual) adherence properties. *Faraday Discuss Chem Soc* 81: 267–80.

Evanson KW, Bannister JP, Leo MD, Jaggar JH. 2014. LRRC26 is a functional BK channel auxiliary γ subunit in arterial smooth muscle cells. *Circ Res* 115: 423–31.

Fuentes S, Bilal U, Galán I, Villalbí JR, Espelt A, Bosque-Prous M, Franco M, Lazo M. 2017. Binge drinking and well-being in European older adults: Do gender and region matter? *Eur J Public Health* 27: 692–9.

Gennis RB. 1989. *Biomembranes: Molecular Structure and Function*. New York: Springer.

Gimpl G, Burger K, Fahrenholz F. 1997. Cholesterol as modulator of receptor function. *Biochemistry* 36: 10959–74.

Granger DN, Rodrigues SF, Yildirim A, Senchenkova EY. 2010. Microvascular responses to cardiovascular risk factors. *Microcirculation* 17: 192–205.

Griguoli M, Sgritta M, Cherubini E. 2016. Presynaptic BK channels control transmitter release: Physiological relevance and potential therapeutic implications. *J Physiol* 594: 3489–500.

Gruner SM. 1985. Intrinsic curvature hypothesis for biomembrane lipid composition: A role for nonbilayer lipids. *Proc Natl Acad Sci USA* 82: 3665–69.

Guimarães S, Lima E, Cipullo J, Lobo S, Burdmann E. 2008. Low insulin-like growth factor-1 and hypocholesterolemia as mortality predictors in acute kidney injury in the intensive care unit. *Crit Care Med* 36: 3165–70.

Hayakawa E, Naganuma M, Mukasa K, Shimozawa T, Araiso T. 1998. Change of motion and localization of cholesterol molecule during L(alpha)-H(II) transition. *Biophys J* 74: 892–8.

Helfrich W. 1973. Elastic properties of lipid bilayers: Theory and possible experiments. *Z Naturforsch C* 28: 693–703.

Helrich CS, Schmucker JA, Woodbury DJ. 2006. Evidence that nystatin channels form at the boundaries, not the interiors of lipid domains. *Biophys J* 91: 1116–27.

Hu G, Antikainen R, Jousilahti P, Kivipelto M, Tuomilehto J. 2008. Total cholesterol and the risk of Parkinson disease. *Neurology* 70: 1972–9.

Huang J, 2009. Model membrane thermodynamics and lateral distribution of cholesterol: from experimental data to Monte Carlo simulation. *Methods Enzymol.* 455: 329–64.

Jaggar JH, Li A, Parfenova H, Liu J, Umstot ES, Dopico AM, Leffler CW. 2005. Heme is a carbon monoxide receptor for large-conductance Ca²⁺-activated K⁺ channels. *Circ Res* 97: 805–12.

Jiang QX. 2019. Cholesterol-dependent gating effects on ion channels. *Adv Exp Med Biol* 1115: 167–90.

Kim JH, London E. 2015. Using sterol substitution to probe the role of membrane domains in membrane functions. *Lipids* 50: 721–34.

Kodama M, Shibata O, Nakamura S, Lee, G. Sugihara. 2004. A monolayer study on three binary mixed systems of dipalmitoyl phosphatidyl choline with cholesterol, cholestanol and stigmasterol. *Colloids Surf B Biointerfaces* 33: 211–26.

Kyle BD, Braun AP. 2014. The regulation of BK channel activity by pre- and post-translational modifications. *Front Physiol* 5: 316.

Latorre R, Castillo K, Carrasquel-Ursulaez W, Sepulveda RV, Gonzalez-Nilo F, Gonzalez C, Alvarez O. 2017. Molecular determinants of BK channel functional diversity and functioning. *Physiol Rev* 97: 39–87.

Le Goff G, Vitha MF, Clarke RJ. 2007. Orientational polarisability of lipid membrane surfaces. *Biochim Biophys Acta* 1768: 562–70.

Lee RM. 1995. Morphology of cerebral arteries. *Pharmacol Ther* 66: 149–73.

Li Q, Yan J. 2016. Modulation of BK channel function by auxiliary beta and gamma subunits. *Int Rev Neurobiol* 128: 51–90.

Lingwood CA. 2011. Glycosphingolipid functions. *Cold Spring Harb Perspect Biol* 3: pii: a004788.

Liu P, Xi Q, Ahmed A, Jaggar JH, Dopico AM. 2004. Essential role for smooth muscle BK channels in alcohol-induced cerebrovascular constriction. *Proc Natl Acad Sci USA* 101: 18217–22.

Loura LM, Prieto M. 1997. Dehydroergosterol structural organization in aqueous medium and in a model system of membranes. *Biophys J* 72: 2226–36.

Lu R, Alioua A, Kumar Y, Eghbali M, Stefani E, Toro L. 2006. MaxiK channel partners: Physiological impact. *J Physiol* 570: 65–72.

Lundbæk, JA. 2008. Lipid bilayer–mediated regulation of ion channel function by amphiphilic drugs. *J Gen Physiol* 131 (5): 421–429.

Lundbæk JA, Andersen OS. 1994. Lysophospholipids modulate channel function by altering the mechanical properties of lipid bilayers. *J Gen Physiol* 104: 645–73.

Lundbæk JA, Birn P, Girshman J, Hansen AJ, Andersen OS. 1996. Membrane stiffness and channel function. *Biochemistry* 35: 3825–30.

Lundbaek JA, Birn P, Hansen AJ, Søgaard R, Nielsen C, Girshman J, Bruno MJ et al. 2004. Regulation of sodium channel function by bilayer elasticity: The importance of hydrophobic coupling: Effects of Micelle-forming amphiphiles and cholesterol. *J Gen Physiol* 123: 599–621.

142   New Techniques for Studying Biomembranes

Lundbæk JA, Collingwood SA, Ingólfsson HI, Kapoor R, Andersen OS. 2010. Lipid bilayer regulation of membrane protein function: Gramicidin channels as molecular force probes. *J R Soc Interface* 7: 373–95.

Mannock DA, McIntosh TJ, Jiang X, Covey DF, McElhaney RN. 2003. Effects of natural and enantiomeric cholesterol on the thermotropic phase behavior and structure of egg sphingomyelin bilayer membranes. *Biophys J* 84: 1038–46.

Marty A, Tan Y, Trautmann A. 1984. Three types of calcium-dependent channel in rat lacrimal glands. *J Physiol* 357: 293–325.

McConnell HM, Radhakrishnan A. 2003. Condensed complexes of cholesterol and phospholipids. *Biochim Biophys Acta* 1610: 159–73.

McIntosh TJ. 1978. The effect of cholesterol on the structure of phosphatidylcholine bilayers. *Biochim Biophys Acta* 513: 43–58.

McMullen TP, Lewis RN, McElhaney RN. 1999. Calorimetric and spectroscopic studies of the effects of cholesterol on the thermotropic phase behavior and organization of a homologous series of linear saturated phosphatidylethanolamine bilayers. *Biochim Biophys Acta* 1416: 119–34.

Meredith AL, Thorneloe KS, Werner ME, Nelson MT, Aldrich RW. 2004. Overactive bladder and incontinence in the absence of the BK large conductance Ca^{2+}-activated K^+ channel. *J Biol Chem* 279: 36746–52.

Miller AA, Budzyn K, Sobey CG. 2010. Vascular dysfunction in cerebrovascular disease: Mechanisms and therapeutic intervention. *Clin Sci (Lond)* 119: 1–17.

Moczydlowski E, Alvarez O, Vergara C, Latorre R. 1985. Effect of phospholipid surface charge on the conductance and gating of a Ca^{2+}-activated K^+ channel in planar lipid bilayers. *J Membr Biol* 83: 273–82.

Needham D, Nunn RS. 1990. Elastic deformation and failure of lipid bilayer membranes containing cholesterol. *Biophys J* 58: 997–1009.

Nezil FA, Bloom M. 1992. Combined influence of cholesterol and synthetic amphiphilic peptides upon bilayer thickness in model membranes. *Biophys J* 61: 1176–83.

North K., Bisen S., Dopico A.M., Bukiya A. 2018. Tyrosine 450 in the BK channel poreforming (slo1) subunit mediates cholesterol protection against alcohol-induced constriction of cerebral arteries. *J Pharmacol Exp Ther* 367: 234–44.

Owen D, Gaus K. 2013. Imaging lipid domains in cell membranes: The advent of superresolution fluorescence microscopy. *Front Plant Sci.* 4: 503.

Peoples RW, Li C, Weight FF. 1996. Lipid vs protein theories of alcohol action in the nervous system. *Annu Rev Pharmacol Toxicol* 36: 185–201.

Pine SH. Steroids. In "Organic chemistry", Misler K.S., Young L.A., eds. McGraw-Hill Book Company, New York, 5th edition, p. 876.

Preston Mason R, Tulenko TN, Jacob RF. 2003. Direct evidence for cholesterol crystalline domains in biological membranes: Role in human pathobiology. *Biochim Biophys Acta* 1610: 198–207.

Róg T, Pasenkiewicz-Gierula M. 2003. Effects of epicholesterol on the phosphatidylcholine bilayer: A molecular simulation study. *Biophys J* 84: 1818–26.

Sackmann E. 1995. Biological membranes architecture and function. In R. Lypowsky and E. Sackmann (Eds.), *Structure and Dynamics of Membranes* (pp. 1–63). Amsterdam, the Netherlands: Elsevier.

Shipston J, Tian L. 2016. Posttranscriptional and posttranslational regulation of BK channels. *Int Rev Neurobiol.* 128: 91–126.

Shrestha N, Bryant S, Thomas C, Richtsmeier D, Pu X, Tinker J, Fologea D. 2017. Stochastic sensing of Angiotensin II with lysenin channels. *Sci Rep.* 7: 2448.

Simakova MN, Bisen S, Dopico AM, Bukiya AN. 2017. Statin therapy exacerbates alcoholinduced constriction of cerebral arteries via modulation of ethanol-induced BK channel inhibition in vascular smooth muscle. *Biochem Pharmacol* 145: 81–93.

Simons K., Sampaio JL. 2011. Membrane organization and lipid rafts. *Cold Spring Harb Perspect Biol.* 3: a004697.

Singh AK, McMillan J, Bukiya AN, Burton B, Parrill AL, Dopico AM. 2012. Multiple cholesterol recognition/interaction amino acid consensus (CRAC) motifs in cytosolic C tail of Slo1 subunit determine cholesterol sensitivity of Ca^{2+}- and voltage-gated K^+ (BK) channels. *J Biol Chem* 287: 20509–21.

Sones WR, Davis AJ, Leblanc N, Greenwood IA. 2010. Cholesterol depletion alters amplitude and pharmacology of vascular calcium-activated chloride channels. *Cardiovasc Res* 87: 476–84.

Stachon A, Böning A, Weisser H, Laczkovics H, Skipka G, Krieg M. 2000. Prognostic significance of low serum cholesterol after cardiothoracic surgery. *Clin Chem* 46: 1114.

Starke-Peterkovic T, Turner N, Else PL, Clarke RJ. 2005. Electric field strength of membrane lipids from vertebrate species: Membrane lipid composition and Na^+-K^+-ATPase molecular activity. *Am J Physiol Regul Integr Comp Physiol* 288: R663–70.

Sugár IP, Chong L.-G. 2012. A statistical mechanical model of cholesterol/phospholipid mixtures: Linking condensed complexes, superlattices, and the phase diagram. *J Am Chem Assoc* 134: 1164–71.

Tao X, Hite RK, MacKinnon R. 2017. Cryo-EM structure of the open high-conductance Ca^{2+}-activated K^+ channel. *Nature* 541: 46–51.

Tormey WP, Moore TM. 2013. Ethanol as a single toxin in non-traumatic deaths: A toxicology perspective. *Leg Med (Tokyo)* 15: 122–5.

Treistman SN, Martin GE. 2009. BK Channels: Mediators and models for alcohol tolerance. *Trends Neurosci* 32: 629–37.

Van Blitterswijk WJ, Van Hoeven RP, Van der Meer BW. 1981. Lipid structural order parameters (reciprocal of fluidity) in biomembranes derived from steady-state fluorescence polarization measurements. *Biochim Biophys Acta* 644: 323–32.

Venegas B, Sugár IP, Chong P-G. 2007. Critical factors for detection of biphasic changes in membrane properties at specific sterol mole fractions for maximal superlattice formation. *J Phys Chem B* 111: 5180–92.

Vyroubala P, Chiarlab C, Giovannini I, Hyspler R, Ticha A, Hrnciarikova D, Zadak Z. 2008. Hypercholesterolemia in clinically serious conditions. *Biomed Pap Med Fac Univ Palacky Olomouc Czech Repub* 152: 181–9.

Wang L, Sigworth FJ. 2009. Structure of the BK potassium channel in a lipid membrane from electron cryomicroscopy. *Nature* 461: 292–5.

Westover EJ, Covey DF. 2004. The enantiomer of cholesterol. *J Membr Biol* 202: 61–72.

Wheeler S, Schmid R, Sillence DJ. 2019. Lipid–Protein interactions in Niemann–Pick Type C disease: Insights from molecular modeling. *Int J Mol Sci* 20: pii: E717.

Wilsnack RW, Wilsnack SC, Kristjanson AF, Vogeltanz-Holm ND, Gmel G. 2009. Gender and alcohol consumption: Patterns from the multinational GENACIS project. *Addiction* 104: 1487–1500.

Wu Y, Liu Y, Hou P, Yan Z, Kong W, Liu B, Li X, Yao J, Zhang Y, Qin F, Ding J. 2013. TRPV1 channels are functionally coupled with BK(mSlo1) channels in rat dorsal root ganglion (DRG) neurons. *PloS One* 8: 10 e78203.

Wu Y, Yang Y, Ye S, Jiang Y. 2010. Structure of the gating ring from the human large-conductance Ca^{2+}-gated K^+ channel. *Nature* 466: 393–7.

Xu X, London E. 2000. The effect of sterol structure on membrane lipid domains reveals how cholesterol can induce lipid domain formation. *Biochemistry* 39: 843–9.

Yuan C, Chen M, Covey DF, Johnston LJ, Treistman SN. 2011. Cholesterol tuning of BK ethanol response is enantioselective, and is a function of accompanying lipids. *PLoS One* 6: e27572.

Yuan C, O'Connell RJ, Jacob RF, Mason RP, Treistman SN. 2007. Regulation of the gating of BKCa channel by lipid bilayer thickness. *J Biol Chem* 282: 7276–86.

Yuan C, O'Connell RJ, Wilson A, Pietrzykowski AZ, Treistman SN. 2008. Acute alcohol tolerance is intrinsic to the BK_{Ca} protein, but is modulated by the lipid environment. *J Biol Chem* 283: 5090–8.

Yuan P, Leonetti MD, Pico AR, Hsiung Y, MacKinnon R. 2010. Structure of the human BK channel Ca^{2+}-activation apparatus at 3.0 A resolution. *Science* 329: 182–6.

Zidovetzki R, Levitan I. 2007. Use of cyclodextrins to manipulate plasma membrane cholesterol content: Evidence, misconceptions and control strategies. *Biochim Biophys Acta* 1768: 1311–24.

8 Restoration of Membrane Environments for Membrane Proteins for Structural and Functional Studies

Liguo Wang

CONTENTS

8.1   INTRODUCTION

Membrane proteins are essential for cellular life. They are involved in various critical biological processes and account for 70% of all known pharmacological targets and 50% of potential new drug targets [1]. As shown by both structural and functional studies, the lipid membrane environment plays an essential role for the structural integrity and activity of membrane proteins [2–7]. Therefore, it is critical to restore the lipid membrane environment of membrane proteins. One method to study membrane proteins in a lipid membrane is to use lipid nanodiscs, where a membrane protein resides in a small patch of lipid bilayer encircled by amphipathic scaffolding proteins [8]. This method has been employed to study the anthrax toxin pore at 22-Å resolution [9], TRPV1 ion channel in complex with ligands at 3–4 Å resolution [10], as well as other membrane proteins [8–18]. Another method, called "random spherically constrained" (RSC) single-particle cryo-EM, where the membrane proteins are reconstituted into liposomes, was developed and employed to study the large conductance voltage- and calcium-activated potassium (BK) channels reconstituted into liposomes at 17-Å resolution [19] and at 3.5-Å resolution [20]. Although both the nanodisc and RSC methods restore the lipid environment of membrane proteins, there is a major difference: the RSC method mimics the cell and provides an asymmetric environment (i.e., inside and outside conditions can be varied independently), whereas there is only one environment surrounding the membrane protein in the nanodisc method. This is especially important for membrane proteins where an asymmetric environment is preferred (e.g., applying ligands to only one side of the membrane or applying transmembrane potential for voltage-gated ion channels or voltage-sensitive proteins or applying pressures to mechanosensitive channels).

After the restoration of lipid membrane environments for membrane proteins, both functions and structures of membrane proteins can be studied. It has been reported that the function of reconstituted potassium channel TRAAK and K2P1 [21,22] was assessed with the pH-sensitive fluorescent dye 9-amino-2-methoxy-6-chloroacridine (ACMA). When K^+ ions flow outward down the K^+ gradient via either a K^+ ionophore or a potassium channel, a negative potential inside liposomes is established. Then H^+ flows down the electrochemical gradient via proton ionophores and quenches the fluorescence signal of ACMA. Here both the polarity and magnitude of the transmembrane potential were tuned by adjusting the chemical gradient across the lipid membrane. The establishment of a negative potential inside liposomes with the desired magnitude was confirmed by the function of the reconstituted hyperpolarization-activated cyclic nucleotide-gated potassium and sodium 2 (HCN2) channels, which open at a voltage around -120 mV [23], opposite to the BK channels (BK channels open at positive transmembrane potentials [24]). After the confirmation of the established transmembrane potential using the flux assay method, cryo-EM was used to study the dipole potential at the center of the lipid bilayer, the transmembrane potential inside liposomes, and the structure of the reconstituted large conductance voltage- and calcium-activated potassium (BK) channel. Using the RSC method, an intermediate state of the BK channel was discovered, which is not accessible by other methods.

8.2   MATERIALS AND METHODS

8.2.1   EXPRESSION AND PURIFICATION OF HCN2 AND BK CHANNELS

The WT HCN2 and the human BK proteins were expressed and purified as previously described [19,25]. Briefly, cells expressing FLAG-tagged HCN2 and BK proteins were broken by sonication for 1 minute; and sonication was repeated twice. Cell debris and nuclei were removed by spinning down at 1,000 g for 15 minutes at 4°C. Supernatants were spun down at 40,000 g for 1 hour at 4°C. Subsequently, pellets were mixed with the extraction buffer (16 mM n-Dodecyl β-D-maltoside (DDM), 200 mM KCl, 50 mM Tris-HCl, pH7.4, 5 mM EDTA, 1x Protease Inhibitor Cocktail (P8340, Sigma)), and rotated for 2 hours at 4°C. Then detergent-resistant membranes were spun down at 17,000 g for 40 minutes at 4°C. Supernatants were collected and mixed with FLAG-beads (anti-FLAG M2 affinity gel A2220, Sigma) and rotated at 4°C for 2 hours for binding. Proteins were eluted with 0.5 mg/mL FLAG peptide (F3290, Sigma) and concentrated.

8.2.2   RECONSTITUTION OF HCN2 CHANNELS FOR FUNCTIONAL STUDIES

The purified HCN2 proteins were reconstituted into liposomes as previously described [19]. The POPC (1-palmitoyl-2-oleoyl-sn-glycero-3-phosphocholine) and POPG (1-palmitoyl-2-oleoyl-sn-glycero-3-phospho-(1′-rac-glycerol)) lipid mixture (3:1 molar ratio) in chloroform was dried under nitrogen for 30 minutes, and rehydrated in buffer A (50 mM Tris-HCl, pH 7.4, 150 mM KCl, 30 mM DM) to a final concentration of 5 mM. The resulted lipid/detergent mixture was mixed with the purified HCN2 channels to a final protein concentration of 0.1 mg/mL (1:5,000 protein: lipid molar ratio). The detergent was removed by serial-dialysis against the buffer containing 20 mM Tris, pH 7.4, 150 mM KCl and 2 mM/0.5 mM/0 mM DM at 4°C for 3 days (one day for each buffer). Empty liposomes were prepared in the same manner without the addition of protein prior to dialysis.

8.2.3   RECONSTITUTION OF BK CHANNELS FOR STRUCTURAL STUDIES

For structural studies, BK channels were reconstituted as previously described [19,26]. Briefly, purified BK channels were mixed with DM-solubilized POPC lipid (Avanti Polar Lipids, Inc.) (DM: POPC = 3:1) giving a final protein-to-lipid molar ratio of 1:1,000. Gel filtration on a hand-packed 24 mL (10 mm I.D and 300 mm height) Sephadex G-50 column was used to remove detergent with a running buffer containing 20 mM HEPES, pH 7.3, 150 mM KCl, 2 mM EDTA. The fractions containing proteoliposomes were collected and concentrated. To obtain highly spherical vesicles, we swelled them by repeated osmotic shocks, adding water to the vesicle suspension (11%, 14%, 18%, 24%, and 33% of the original volume) at 1 hour intervals at 4°C.

8.2.4   VESICLE FORMATION FOR THE MEASUREMENT OF DIPOLE
POTENTIALS AT THE BILAYER CENTER

Diphytanoyl phosphatidylcholine (ester-DPhPC) and diphytanyl phosphatidylcholine (ether-DPhPC) (Avanti Polar Lipids, Inc.) were used as received. The lipids were hydrated in HEPES-buffered KCl solution (135 mM KCl, 5 mM NaCl, 1 mM EDTA,

10 mM HEPES, pH = 7.4) to a concentration of 4 mg/mL, frozen and thawed 10 times, and extruded through an 80-nm polycarbonate membrane filter (Whatman) using a LipexTM extruder (Northern Lipids Inc.) [27]. To obtain highly spherical vesicles, we swelled them by repeated osmotic shocks, adding water to the vesicle suspension (11%, 14%, 18%, 24%, and 33% of the original volume) at 1 hour intervals at room temperature.

8.2.5 FLUORESCENCE BASED FLUX ASSAY FOR FUNCTIONAL STUDIES OF RECONSTITUTED HCN2 CHANNELS

Empty liposomes or proteoliposome were 100-fold diluted into a buffer containing 150 mM NaCl, 20 mM Tris-HCl, pH 7.4, and 2 μM 9-amino-2-methoxy-6-chloroacridine (ACMA, Sigma-Aldrich). This yielded a final K^+ concentration of 1.5 mM outside liposomes. Fluorescence intensity was measured every 1 second using a SpectraMax fluorometer (Molecular Devices) for a total of 180 seconds with excitation at 395 nm and emission at 490 nm. The proton ionophore carbonyl cyanide m-chlorophenyl hydrazone (CCCP, Sigma-Aldrich) was added to a final concentration of 1 μM after 60 seconds, and the sample was gently mixed with a pipette. Valinomycin (Sigma-Aldrich) was added to a final concentration of 20 nM at 120 seconds.

8.2.6 CRYO-SAMPLE PREPARATION AND IMAGING FOR DIPOLE POTENTIAL MEASUREMENT

The cryo-EM samples for the dipole potential measurement were prepared as previously described [28]. Six microliters of the swollen liposome solution were applied to the front side of a glow-discharged holey carbon grid Quantifoil R2/2 (Quantifoil, Germany) and blotted from the back side for 3–6 seconds with a slip of filter paper (Whatman). The specimen was rapidly frozen by plunging into liquid ethane and stored in liquid nitrogen. Images of liposomes within the holes in the carbon film were obtained using a Tecnai F20 electron microscope at 200 keV using 20 or 30 μm objective apertures. The dose for each exposure was about 20 e/Å^2. Images were taken at 45,000 or 50,000 magnification and −2.0 to −3.3 μm defocus and recorded on Kodak SO-163 film. This level of defocus was chosen because in simulations a defocus of −2.5 μm gave the largest sensitivity to variations of electron scattering in the membrane interior. Negatives were scanned with a Zeiss SCAI film scanner to an effective pixel size of 2.4 Å. Estimates of the defocus and other parameters of the contrast-transfer function were obtained by fits-to-image power spectra from the carbon surrounding the holes, under the assumption that the amorphous carbon is a random object with constant structure factor magnitudes in the spatial frequency range of 1/30–1/10 Å^{-1}.

8.2.7 CRYO-SAMPLE PREPARATION AND IMAGING OF BK PROTEOLIPOSOMES

BK proteoliposomes were swollen right before frozen by repeated osmotic shocks as described in Section 2.3. Then 2 μL of BK proteoliposomes at a concentration of 2 mM lipid and 1 mg/mL protein was applied onto a glow-discharged perforated carbon grids (CFlat R2/2, EMS), and incubated for 5–10 minutes at 22°C and 100% relative humidity in an FEI Vitrobot Mark IV (FEI). After incubation, the sample was blotted from the edge and 2 μL of buffer containing 10 mM HEPES, pH 7.3, 75 mM KCl, and 1 mM EDTA was applied onto the TEM grid to swell proteoliposomes. Then grids were blotted for 7 seconds with force 0 and fast frozen in liquid ethane cooled by liquid nitrogen. Grids were then transferred to an FEI Titan Krios electron microscope operating at an acceleration voltage of 300 keV with an energy filter (20 eV slit width). Images were recorded in an automated fashion on a Gatan K2 Summit (Gatan) detector set to super-resolution mode with a super-resolution pixel size of 0.525 Å using Leginon [29]. The dose rate on the camera was set to be ~8 counts per physical pixel per second. The total exposure time was 8.6 seconds (0.2 second/frame), leading to a total accumulated dose of 60 e$^-$/Å^2 on the specimen.

8.2.8 LIPOSOME SUBTRACTION

To remove the membrane contribution from the recorded proteoliposome images, a scalable model of a liposome was built using an electron scattering profile of the phospholipid bilayer as previously described [19,26,28,30]. The electron scattering profile can be derived from molecular dynamics simulations or determined empirically [28] by the Hankel transform [31] as shown in Figure 8.1a. First, a 3D model of

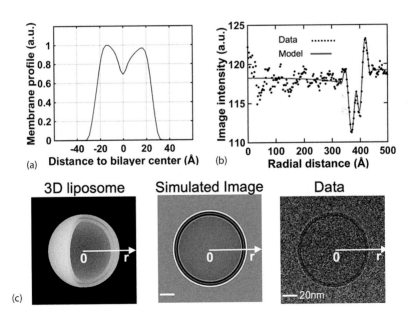

FIGURE 8.1 Liposome subtraction. (a) Membrane scattering profile of a POPC bilayer. (b) Comparison of the circularly averaged image intensity of an experimental cryo-EM image shown in (c) and that of a simulated cryo-EM image shown in (c). (c) Modeling of a liposome using the membrane profile shown in (a).

a liposome is built using the membrane profile (Figure 8.1c). Then the 3D liposome is projected to the x-y plane and convoluted with the contrast transfer function (CTF) of the microscope [32]. Finally, the simulated image is scaled by a scaling factor determined using least squares method to match the experimental liposome image and subtracted from the experimental liposome image. As shown in Figure 8.1b, the simulated image agrees well with the experimental image.

8.3   RESULTS

8.3.1   ESTABLISHMENT OF A NEGATIVE TRANSMEMBRANE POTENTIAL TO OPEN RECONSTITUTED HCN2 CHANNELS

In nature, membrane proteins are in an asymmetric environment and ligands or voltages on one side of the membrane differ from those on the other side. The RSC method can provide an asymmetric environment similar to that in a cell (internal and external conditions can be varied independently). Additionally, this asymmetry can be evaluated experimentally. For example, when K^+ ions flow out of liposomes via valinomycin (K^+ ionophore), a negative potential is established inside liposomes (Figure 8.2a). The H^+ ions flow down the electrochemical gradient (entering liposomes) via the proton ionophore (CCCP), and quenches the fluorescence signal of the pH sensitive dye (ACMA).

To estimate the magnitude of the established negative potential, HCN2 channels were reconstituted into liposomes. At a negative transmembrane potential, HCN2 channels open and allow both K^+ and Na^+ to cross the membrane (for reviews see [33,34]). As shown in Figure 8.2a, the only positive ion (K^+) crosses the membrane via K^+ ionophore (Valinomycin), which results in a negative potential. Whether the HCN2 channels open or not does not change the negative potential. However, when Na^+ ions are present (Figure 8.2b), HCN channels open and allow Na^+ ions to enter liposomes. The influx of Na^+ ions counterbalances the efflux of K^+ ions, resulting in a zero transmembrane potential. The ACMA quenching should decrease, and it did (Figure 8.2c). The reason that not all ACMA quenching was removed was due to the presence of empty liposomes and liposomes containing inside-out inserted HCN2 channels which were not open. When the transmembrane potential was changed from -120 to -90 mV and -60 mV, which were higher than the $V_{1/2}$ of mouse HCN2 expressed in HEK293 cell (-104 mV) [35], only a small fraction of HCN2 channels should open or no HCN2 channels should open at all. Thus, there should be no difference in ACMA quenching between the experiments with and without Na^+ ions. Our results confirmed that (Figure 8.2d). This demonstrated that the absolute magnitude of the transmembrane potential can be established as designed.

As shown in Figure 8.2c and d, a negative potential inside liposomes could be established to open HCN2 channels. However, there is one question to be addressed: how long could the transmembrane potential hold (i.e., are liposomes leaky)? To assess the breakage of liposomes during the exchange of external solutions to establish the K^+ gradient across the membrane or the leakage during subsequent measurements, extruded liposomes were loaded with a non-membrane-permeant

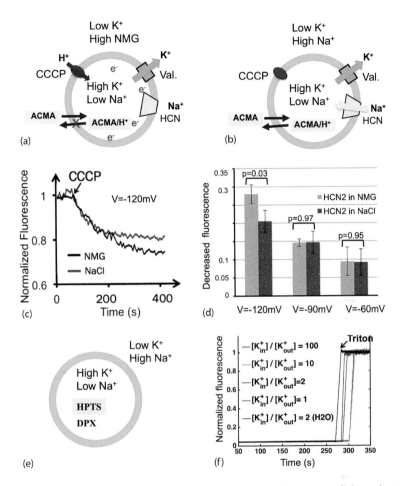

FIGURE 8.2 Establishment of a negative potential inside liposomes. Schematic of the flux assay in the absence (a) and presence (b) of Na$^+$ (b). The efflux of K$^+$ results in a negative transmembrane potential (a), and the influx of Na$^+$ counterbalances the efflux of K$^+$ (b). (c) Representative fluorescence profiles with a 100-fold [K$^+$]$_{in}$/[K$^+$]$_{out}$ gradient (-120 mV). (d) Amount of ACMA quenching with and without Na$^+$ at different transmembrane potentials, which were calculated using the Nernst equation. (e) Schematic diagram of the leakage assay. (f) The fluorescence did not change during the dilution of liposomes and the subsequent measurement. At $\sim$280 s, Triton X-100 (0.1%) was added to solubilize liposomes for complete release of HPTS from liposomes.

fluorescence dye (HPTS) and its quencher (DPX) as shown in Figure 8.2e. With the exchange of external solutions from high K$^+$ to low K$^+$ concentration, there was no observable release of HPTS from liposomes (Figure 8.2f) during the dilution of liposomes and the subsequent measurement. This confirms that liposomes remain intact (not broken) during the change of solutions to build the chemical gradient across the membrane and do not leak after the establishment of the chemical gradient across the membrane. This was also supported by our previous observation that the fluorescence signal of a voltage sensitive dye in BK proteo-liposomes remained constant for at least 10 minutes after the system reached an equilibrium [19].

8.3.2 MEASUREMENT OF THE DIPOLE POTENTIAL IN LIPID MEMBRANES

Besides the transmembrane potential, there is a dipole potential, a membrane-internal potential from the dipolar components of the phospholipids and interface water [36]. The dipole potential of a lipid bilayer membrane accounts for its much larger permeability to anions than cations, and affects the conformation and function of membrane proteins. The absolute value of the dipole potential has been very difficult to measure, although its value has been estimated to be in the range of 200–1000 mV from ion translocation rates, the surface potential of lipid monolayers, and molecular dynamics (MD) calculations [37–41]. Here, cryo-EM method was employed to investigate the dipole potentials of both ester and ether lipid membranes [42].

In cryo-EM, the primary mechanism for image contrast is the phase shift in the electron wave function (elastic scattering) as it passes through the specimen. The total phase shift is proportional to the projected potential, that is the integrated electrostatic potential along the path of the electron. The intensity of the recorded image is expected to vary in proportion to the phase shift when the weak phase object approximation is employed for defocused imaging [43]. Thus, the recorded image intensity is a reflection of the projected potential of the specimen, the superposition of the atomic potential and any additional electrostatic potentials.

To estimate the dipole potential, the membrane profile was obtained from each individual liposome image based on the Fourier slice theorem [44–46]. Images of liposomes were corrected for the microscope contrast-transfer function (CTF) [32], rotationally averaged, and an inverse Abel transform yields the membrane profile (i.e., the phase-shift profile) as a function of the distance z normal to the membrane plane (Figure 8.3a–d). The atomic potential profile across the membrane is computed from the atom density obtained from MD simulations [47] as previously described [28]. After the subtraction of atomic potential from the membrane profile, the dipole potential profile across the membrane was obtained (Figure 8.3e). The peak dipole potential was estimated to be 510 and 260 mV for ester-DPhPC and ether-DPhPC (Figure 8.3e–h), respectively. The dipole potential of ester DPhPC was 228 mV measured using bilayer method [48] and 1002 mV measured in MD simulations [47]. The dipole potential of ether DPhPC was estimated to be 567 mV in MD simulations [47]. The cryo-EM value is smaller than that in MD simulations, but larger than that in the planar lipid bilayer measurements. However, the ratio between the ester and ether lipids is the same as those observed in both MD simulations and planar lipid bilayer and monolayer measurements for lipids including DPhPC [28,36].

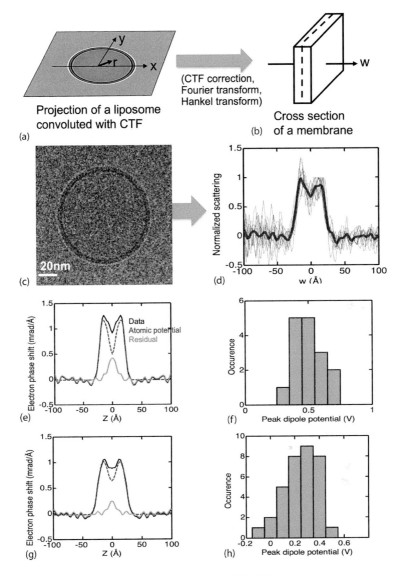

FIGURE 8.3 Dipole potential in ester and ether DPhPC bilayers. (a) A simulated cryo-EM image of a liposome. (b) A planar lipid bilayer. (c) An experimental cryo-EM image of a liposome. (d) Individual membrane profiles are shown aligned and superimposed, with the mean indicated in red. (e) Model fitting of ester DPhPC. The solid black curve is the averaged and symmetrized membrane profile. The red curve is the atomic phase shift profile, and the green curve is the residual, taken to be the dipole potential phase shift. (f) Histogram of peak dipole potentials obtained from fitting models to the image densities from 16 liposomes. (g) Model fitting of ether DPhPC. The solid black curve is the averaged and symmetrized membrane profile. The red curve is the atomic phase shift profile, and the green curve is the residual, taken to be the dipole potential phase shift. (h) Histogram of peak dipole potentials obtained from fitting models to the image densities from 34 liposomes.

8.3.3 OBSERVATION OF A POSITIVE TRANSMEMBRANE POTENTIAL BY CRYO-EM

As shown in Section 3.1, a negative transmembrane potential can be established with the desired magnitude. It would be ideal if all the reconstituted HCN2 channels were inserted outside-out. However, only about 20% of the reconstituted membrane proteins are inserted outside-out (the rest are inserted inside-out) [19]. To open the majority of HCN2 channels (the inside-out inserted channels), the potential inside liposomes needs to be positive with respect to the external solution. As shown in Figure 8.4a, a positive transmembrane potential can be established. However, this positive transmembrane potential cannot be assessed using the flux assay method discussed in Section 3.1. Here, cryo-EM method was employed. The potential associated with a liposome made of neutral lipids consists of a dipole potential and a transmembrane potential (Figure 8.4b). The corresponding membrane scattering profile is asymmetric as shown in Figure 8.4c.

To use the cryo-EM method to assess the transmembrane potential, cryo-EM images of liposomes were circularly averaged and compared with that of a simulated

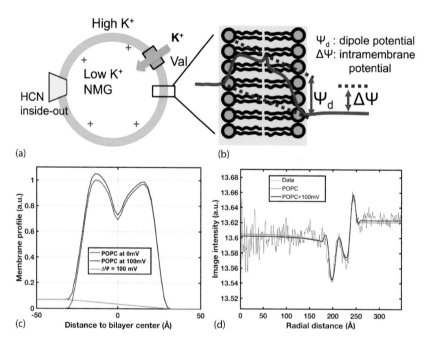

FIGURE 8.4 Measurement of transmembrane potential using cryo-EM. (a) Schematic diagram to obtain a positive transmembrane potential inside liposomes. (b) Transmembrane potential ($\Delta\Psi$) and dipole potential (Ψd) in a lipid membrane. (c) Membrane profile of POPC bilayer in the presence of a positive transmembrane potential. (d) Circularly averaged experimental image intensity (red curve) is compared with the POPC membrane profile in the absence (green curve) and presence (blue curve) of a +100 mV transmembrane potential, respectively.

liposome. As shown in Figure 8.4d, the membrane profile including a positive transmembrane potential fits the experimental profile better than that without the positive transmembrane potential. As the effect of the transmembrane potential on the circularly averaged liposome intensity is less obvious compared with that of the dipole potential (Figures 8.3e and 8.4c), the accuracy of the transmembrane potential measured using this method is less than that of the dipole potential measurement. Another way to assess the transmembrane potential is to study the effect of the transmembrane potential on the structure of the HCN2 channel using the RSC method. As shown in Section 3.1, HCN2 channels open at -120 mV and close at -90 mV. If the structure of HCN channel at -120 mV is different from that at -90 mV, then the magnitude of the transmembrane potential is confirmed (the inside-out inserted HCN2 channels experience a negative transmembrane potential with the setting shown in Figure 8.4a).

8.3.4 An Intermediate State of the BK Channel Reconstituted in Liposomes

In the RSC method, membrane proteins are reconstituted into liposomes, and an asymmetric environment is restored for membrane proteins. This is essential for studies of membrane proteins, where an asymmetric environment is preferred (e.g., applying ligands to only one side of the membrane or applying a transmembrane potential for voltage-gated ion channels or voltage-sensitive proteins or applying pressures to mechanosensitive channels). As a proof of concept, the structure of a BK channel at zero transmembrane potential was studied. After the collection of cryo-EM images of BK proteoliposomes (Figure 8.5a), the liposome contribution was removed (Figure 8.5b). The BK protein particles were picked from the liposome-subtracted images (Figure 8.5b). The class averages of the dataset clearly show different views of the BK channel and the presence of the lipid bilayer (Figure 8.5c and d). The position of lipid bilayer can be seen clearly (Figure 8.5c–f). Using the liposome-subtracted protein particle images, a 3D cryo-EM structure was determined [20] (Figure 8.5e and f).

The BK channel is a member of the *Slo* family, whose α-subunits contain regulator-of-conductance-for-K^+ (RCK) domains in their large intracellular C-terminal region. In all determined structures made of RCK domains [49–56], a four-fold symmetry along the central pore (C4) was observed. With the restoration of the lipid membrane of the BK channel, a two-fold symmetry (C2) was observed in the 3.5-Å EM density map of the BK channel [20]. This is an intermediate state of the BK channel and has not been accessible by other methods (e.g., x-ray crystallography, cryo-EM of detergent solubilized membrane proteins). This two-fold symmetry in a homotetramer channel was also observed in the recently determined structure of a voltage-gated sodium channel (the human Nav1.7 channel) [57]. Thus, it is critical to reconstitute membrane proteins into liposomes in order to probe more physiologically relevant functional states of membrane proteins.

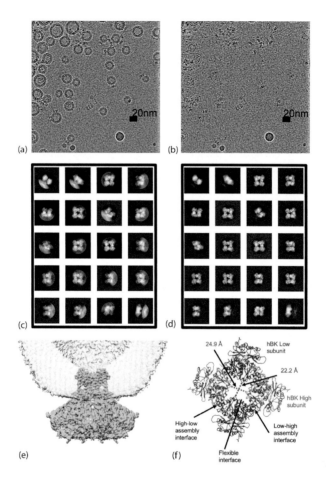

FIGURE 8.5 Study of BK channels reconstituted in POPC liposomes. (a) A representative cryo-EM image of BK proteoliposomes at −3.8 μm defocus. (b) Liposomes are subtracted from the image shown A. BK particles are marked with red boxes (15 nm). (c,d) The largest 20 2D-class averages of 122,000 BK particles before (c) and after (d) liposome subtraction. Box size is 27 nm, and the circular mask is 17 nm in diameter. (e) BK cryo-EM density map. Lipid membrane is shown in gray mesh. For clarity, only half of the lipid membrane is shown. (f) View of the gating ring in hBK from the extracellular side. The diagonal distances between the Cα atoms of Val 785 are indicated.

8.4 SUMMARY

To study functions and structures of voltage-gated ion channels and voltage-sensitive membrane proteins *in vitro*, the restoration of lipid membrane environment and the manipulation of transmembrane potentials are critical. In the RSC method, membrane proteins are reconstituted into liposomes and the transmembrane potential can be tuned by adjusting the chemical gradient across the membrane. The establishment of a transmembrane potential was confirmed by the functional assay of the HCN channels and the transmembrane potential could be held for a long period of time. The RSC method was successfully employed to study the structure of a voltage-gated K^+ channel (the BK channel) and an intermediate state was discovered, which is not accessible by other methods, suggesting the importance of the lipid environment.

REFERENCES

1. Weinglass, A.B., J.P. Whitelegge, and H.R. Kaback, Integrating mass spectrometry into membrane protein drug discovery. *Current Opinion In Drug Discovery & Development*, 2004. **7**(5): pp. 589–599.
2. Schmidt, D., Q.X. Jiang, and R. MacKinnon, Phospholipids and the origin of cationic gating charges in voltage sensors. *Nature*, 2006. **444**(7120): pp. 775–779.
3. Gonen, T., et al., Lipid-protein interactions in double-layered two-dimensional AQP0 crystals. *Nature*, 2005. **438**(7068): pp. 633–638.
4. Long, S.B., et al., Atomic structure of a voltage-dependent K^+ channel in a lipid membrane-like environment. *Nature*, 2007. **450**(7168): pp. 376–383.
5. Hilgemann, D.W., Getting ready for the decade of the lipids. *Annual Review of Physiology*, 2003. **65**: pp. 697–700.
6. Hille, B., et al., Phosphoinositides regulate ion channels. *Biochimica et Biophysica Acta (BBA) - Molecular and Cell Biology of Lipids*, 2015. **1851**(6): pp. 844–856.
7. Lee, A.G., Biological membranes: The importance of molecular detail. *Trends in Biochemical Sciences*, 2011. **36**(9): pp. 493–500.
8. Bayburt, T.H., Y.V. Grinkova, and S.G. Sligar, Self-assembly of discoidal phospholipid bilayer nanoparticles with membrane scaffold proteins. *Nano Letters*, 2002. **2**(8): pp. 853–856.
9. Katayama, H., et al., Three-dimensional structure of the anthrax toxin pore inserted into lipid nanodiscs and lipid vesicles. *Proceedings of the National Academy of Sciences of the United States of America*, 2010. **107**(8): pp. 3453–3457.
10. Gao, Y., et al., TRPV1 structures in nanodiscs reveal mechanisms of ligand and lipid action. *Nature*, 2016. **534**(7607): pp. 347–351.
11. Jackson, S.M., et al., Structural basis of small-molecule inhibition of human multidrug transporter ABCG2. *Nature Structural & Molecular Biology*, 2018. **25**(4): pp. 333–340.
12. Taylor, N.M.I., et al., Structure of the human multidrug transporter ABCG2. *Nature*, 2017. **546**(7659): pp. 504–509.
13. Srivastava, A.P., et al., High-resolution cryo-EM analysis of the yeast ATP synthase in a lipid membrane. *Science*, 2018. **360**(6389): p. eaas9699.
14. Roh, S.-H., et al., The 3.5-Å CryoEM structure of nanodisc-reconstituted yeast vacuolar ATPase vo proton channel. *Molecular Cell*, 2018. **69**(6): pp. 993–1004.e3.
15. Dang, S., et al., Cryo-EM structures of the TMEM16A calcium-activated chloride channel. *Nature*, 2017. **552**(7685): pp. 426–429.
16. McGoldrick, L.L., et al., Opening of the human epithelial calcium channel TRPV6. *Nature*, 2018. **553**(7687): pp. 233–237.
17. Autzen, H.E., et al., Structure of the human TRPM4 ion channel in a lipid nanodisc. *Science*, 2018. **359**(6372): pp. 228–232.
18. Chen, Q., et al., Structure of mammalian endolysosomal TRPML1 channel in nanodiscs. *Nature*, 2017. **550**(7676): pp. 415–418.
19. Wang, L. and F.J. Sigworth, Structure of the BK potassium channel in a lipid membrane from electron cryomicroscopy. *Nature*, 2009. **461**(7261): pp. 292–295.
20. Tonggu, L. and L. Wang, Broken symmetry in the human BK channel. *bioRxiv*, 2018: p. 494385.

21. Brohawn, S.G., J. del Marmol, and R. MacKinnon, Crystal structure of the human K2P TRAAK, a lipid- and mechano-sensitive K$^+$ ion channel. *Science*, 2012. **335**(6067): pp. 436–441.

22. Miller, A.N. and S.B. Long, Crystal structure of the human two-pore domain potassium channel K2P1. *Science*, 2012. **335**(6067): pp. 432–436.

23. Qu, J., et al., Functional comparison of HCN isoforms expressed in ventricular and HEK 293 cells. *Pflugers Archiv European Journal of Physiology*, 2002. **444**(5): pp. 597–601.

24. Horrigan, F.T. and R.W. Aldrich, Coupling between voltage sensor activation, Ca^{2+} binding and channel opening in large conductance (BK) potassium channels. *Journal of General Physiology*, 2002. **120**(3): pp. 267–305.

25. Li, M., et al., Effects of N-glycosylation on hyperpolarization-activated cyclic nucleotide-gated (HCN) channels. *Biochemical Journal*, 2015. **466**: pp. 77–84.

26. Wang, L., Random spherically constrained single-particle (RSC) method to study voltage-gated ion channels. *Methods in Molecular Biology*, 2018. **1684**: pp. 265–277.

27. Mayer, L.D., M.J. Hope, and P.R. Cullis, Vesicles of variable sizes produced by a rapid extrusion procedure. *Biochimica Et Biophysica Acta*, 1986. **858**(1): pp. 161–168.

28. Wang, L., P.S. Bose, and F.J. Sigworth, Using cryo-EM to measure the dipole potential of a lipid membrane. *Proceedings of the National Academy of Sciences of the United States of America*, 2006. **103**(49): pp. 18528–18533.

29. Suloway, C., et al., Automated molecular microscopy: The new Leginon system. *Journal of Structural Biology*, 2005. **151**(1): pp. 41–60.

30. Jiang, Q.X., D.W. Chester, and F.J. Sigworth, Spherical reconstruction: A method for structure determination of membrane proteins from cryo-EM images. *Journal of Structural Biology*, 2001. **133**(2–3): pp. 119–131.

31. Bracewell, R.N., *The Fourier Transform and its Applications*. 3rd ed. 2000, Boston, MA: McGraw-Hill.

32. Frank, J., *Three-dimensional Electron Microscopy of Macromolecular Assemblies*. 2nd ed. 2006, New York: Oxford University Press.

33. Lewis, A.S., C.M. Estep, and D.M. Chetkovich, The fast and slow ups and downs of HCN channel regulation. *Channels*, 2010. **4**(3): 215–231.

34. Robinson, R.B. and S.A. Siegelbaum, Hyperpolarization-activated cation currents: From molecules to physiological function. *Annu Rev Physiol*, 2003. **65**: pp. 453–480.

35. Ye, B. and J.M. Nerbonne, Proteolytic processing of HCN2 and co-assembly with HCN4 in the generation of cardiac pacemaker channels. *J Biol Chem*, 2009. **284**(38): pp. 25553–25559.

36. Wang, L., Measurements and implications of the membrane dipole potential. *Ann Rev Biochem*, 2012. **81**(1): pp. 615–635.

37. Brockman, H., Dipole potential of lipid membranes. *Chem Phys Lipids*, 1994. **73**(1–2): pp. 57–79.

38. Clarke, R.J., The dipole potential of phospholipid membranes and methods for its detection. *Adv Colloid Interf Sci*, 2001. **89–90**: pp. 263–281.

39. Pohl, E.E., Dipole potential of bilayer membranes, in *Advances in Planar Lipid Bilayers and Liposomes*, H.T. Tien and A. Ottova-Leitmannova, Eds., 2005, Cambridge, MA: Academic Press. pp. 77–100.

40. Poignard, C., et al., Ion fluxes, transmembrane potential, and osmotic stabilization: A new dynamic electrophysiological model for eukaryotic cells. *Europ Biophys J*, 2011. **40**(3): pp. 235–246.

41. Honig, B.H., W.L. Hubbell, and R.F. Flewelling, Electrostatic interactions in membranes and proteins. *Ann Rev Biophy Biophys Chem*, 1986. **15**(1): pp. 163–193.

42. Jiang, Y.Q., et al., Characteristics of HCN channels and their participation in neuropathic pain. *Neurochem Res*, 2008. **33**(10): pp. 1979–1989.

43. Kirkland, E.J., *Advanced Computing in Electron Microscopy.* 1998, New York: Plenum Press.

44. Stallmeyer, M.J.B., et al., Image reconstruction of the flagellar basal body of *Salmonella typhimurium. J Mol Biol*, 1989. **205**(3): pp. 519–528.

45. Sosinsky, G.E., et al., Substructure of the flagellar basal body of Salmonella typhimurium. *J Mol Biol*, 1992. **223**(1): pp. 171–184.

46. Francis, N.R., et al., Isolation, characterization and structure of bacterial flagellar motors containing the switch complex. *J Mol Biol*, 1994. **235**(4): pp. 1261.

47. Shinoda, K., et al., Comparative molecular dynamics study of ether- and ester-linked phospholipid bilayers. *J Chem Phys*, 2004. **121**(19): pp. 9648–9654.

48. Peterson, U., et al., Origin of membrane dipole potential: Contribution of the phospholipid fatty acid chains. *Chem Phys Lip*, 2002. **117**(1–2): pp. 19–27.

49. Jiang, Y., et al., Structure of the RCK domain from the *E. coli* K^+ channel and demonstration of its presence in the human BK channel. *Neuron*, 2001. **29**(3): pp. 593–601.

50. Jiang, Y.X., et al., Crystal structure and mechanism of a calcium-gated potassium channel. *Nature*, 2002. **417**(6888): pp. 515–522.

51. Jiang, Y.X., et al., The open pore conformation of potassium channels. *Nature*, 2002. **417**(6888): pp. 523–526.

52. Wu, Y., et al., Structure of the gating ring from the human large-conductance Ca^{2+}-gated K^+ channel. *Nature*, 2010. **466**(7304): pp. 393–397.

53. Yuan, P., et al., Structure of the human BK channel Ca^{2+}-activation apparatus at 3.0 Å resolution. *Science*, 2010. **329**(5988): pp. 182–186.

54. Yuan, P., et al., Open structure of the Ca^{2+} gating ring in the high-conductance Ca^{2+}-activated K^+ channel. *Nature*, 2012. **481**(7379): pp. 94–97.

55. Hite, R.K., X. Tao, and R. MacKinnon, Structural basis for gating the high-conductance Ca^{2+}-activated K^+ channel. *Nature*, 2017. **541**(7635): pp. 52–57.

56. Tao, X., R.K. Hite, and R. MacKinnon, Cryo-EM structure of the open high-conductance Ca^{2+}-activated K^+ channel. *Nature*, 2017. **541**(7635): pp. 46–51.

57. Xu, H., et al., Structural basis of Nav1.7 Inhibition by a gating-modifier spider toxin. *Cell*, 2019. **176**(4): pp. 702–715.

Index

Note: Page numbers in italic and bold refer to figures and tables, respectively.

Methods in Signal Transduction

Series Editors: Joseph Eichberg Jr., Michael X. Zhu, and Harpreet Singh

Previously Published Volumes

G Protein-Coupled Receptors
Tatsuya Haga and Gabriel Berstein

Signaling through Cell Adhesion Molecules
Jun-Lin Guan